The Medical Science Liaison: An A to Z Guide

A book by
ERIN ALBERT WITH CATHLEEN SASS

AuthorHouse™
1663 Liberty Drive, Suite 200
Bloomington, IN 47403
www.authorhouse.com
Phone: 1-800-839-8640

© 2007 Pharm, LLC. All rights reserved.

No part of this publication may be reproduced, stored in a retrieval system, or transmitted in any form or by any means, electronic, mechanical, photocopying, recording, scanning, or otherwise, except as permitted under Section 107 or 108 of the 1976 United States Copyright Act, without the prior written permission of the CEO of Pharm, LLC.

Limit of Liability/Disclaimer of Warranty: While the publisher and author have used their best efforts in preparing this book, they make no representations or warranties with respect to the accuracy or completeness of the contents of this book and specifically disclaim any implied warranties of merchantability or fitness for a particular purpose. No warranty may be created or extended by sales representatives or written sales materials. The advice and strategies contained herein may not be suitable for your situation. You should consult with a professional where appropriate. Neither the publisher nor author shall be liable for any loss of profit or any other commercial damages, including but not limited to special, incidental, consequential, or other damages. The information contained herein is not necessarily the opinion of the author or publisher.

Designations used by companies to distinguish their products are often claimed by trademarks. In all instances where the author or publisher is aware of a claim, the trademarks have been noted where applicable. The inclusion of a trademark does not imply an endorsement or judgment of a product or service of another company, nor does it imply an endorsement or judgment by another company of this book or the opinions contained herein.

First published by AuthorHouse 10/1/2007

ISBN: 978-1-4343-3750-4 (sc)

Library of Congress Control Number: 2007907127

Printed in the United States of America
Bloomington, Indiana

This book is printed on acid-free paper.

*To my mom, Dorothy, dad, John, and brother, Mark -
for always reminding me that life is not a dress rehearsal*
—*Erin*

In Memory of Mac Williamson
—*Cathy*

Table of Contents

Introduction: The History of the MSL ... 1
   The Past and Future of MSLs:
      Interview with Stan Bernard, MD, MBA ... 2

Part I: Becoming a MSL

Chapter 1: What is a MSL? ... 9
   The Real World: An Interview with a Working MSL -
      Ernie Pitting, RPh, MS, PharmD candidate ... 24

Chapter 2: Activities, Salary, Job
   Satisfaction and Career Paths of the MSL ... 31
   Benefits for the MSL: Interview with Chris Conley, CSAM ... 41

Chapter 3: Getting Hired ... 49
   Getting Into the MSL Role:
      A Recruiter's Perspective - Interview with Tony Beachler ... 75

Part II: Perfecting the Art of Liaising

Chapter 4: Starting off as a MSL ... 85
   Legal/Regulatory Issues for the MSL To Consider:
      Interview with Christopher R. Hall, Esq. ... 96
   The International MSL Perspective:
      Interview with Michael Hamann, PhD ... 103

Chapter 5: Customers' Viewpoints ... 107
   Interview with Chris Bojrab, MD ... 107
   Interview with Ron Chervin, MD, MS ... 111
   Interview with James A. Simon, MD, CCD, FACOG ... 115
   Interview with Paul Keck, MD ... 120

Chapter 6: Research & Technology ... 125
   Resources For the MSL: An Interview with LouAnn Fare, MS ... 125
   Managing Medical Information: An Interview with Dr. Amy Peak ... 130
   About Metrics: An Interview with Michael Taylor ... 137

Chapter 7: Work + Life = Balance? 145
    The MSL Father: Interview with Tim Hill, PharmD, MS 149
    Working as a MSL/Mom: Interview with Carole Carter-Olkowski 154

Part III: Post MSL - What's Next?

Chapter 8: Job Hopping 165
    The MSL Moving Onward: Interview with Bryan Vaughn 165

Chapter 9: Management of MSLs 175
    Medical Science Liaison Outsourcing:
        Interview with Kyle Kennedy 175
    Interview with a MSL Manager:
        Susan E. Malecha, PharmD, MBA 183

Chapter 10: The Afterlife 189
    Academic Pharmacy After the MSL role:
        Interview with Scott Stolte, PharmD 192
    Entrepreneurialism Post MSL: Interview with Jane Chin, PhD 196
    Life After the MSL Role in Medical Education:
        Interview with Matt Lewis, MPA 200

Chapter 11: The Future of the MSL Role 207
    The Clinical Trial Liaison: Interview with Brian Best 207
    Future Of The MSL: Closing Interview with Dr. Stan Bernard 214

Appendix A:
    Major Medical Associations/Meetings for Therapeutic Areas 217

Appendix B:
    Real World Career Paths to and Beyond the MSL 225

Appendix C:
    Acronyms & Abbreviations 237

Bibliography, References, and Resources 245

Acknowledgements 249

Introduction: The History of the MSL

Many know that the origin of the Medical Science Liaison (MSL) came out of Upjohn. What is not largely known, however, is the drug that started the MSL movement: it was tolbutamide (Orinase®). According to the text, *A Century of Caring: The Upjohn Story* by Robert Carlisle, the first ever MSL program was initiated by marketing to establish rapport and provide educational awareness to doctors *before* they were able to write prescriptions, during medical school. In 1967, the program was started and seeded by sales professionals that had, "intense interest in science, [had] high social skills, and recognize[d] that they [were] no longer detailing products".[1]

The MSLs began calling on medical schools and thus, a movement was born. Upjohn had been mostly unknown to medical students, but after just 6 years of the MSL program, "Upjohn was one of the best-known"[1] companies to medical students. The entire onus of the MSL was to provide education in the form of monographs, access to internal researchers, and assistance for external educational programs. The program was conceptualized and developed a mission to provide appropriate use of oral diabetic drugs, which at the time was a novel approach to diabetes.[1]

Today, there are thousands of MSLs across the country, in hundreds of disease states and therapeutic areas. As will be explored further in this book, MSL job descriptions, professional backgrounds, and job motivators can vary widely from company to company. This text will discuss who the MSL is, what the MSL does, and where the role might be headed in the future. The approach we are taking with the book is to look at the product life cycle of the MSL, so to speak. We begin with a professional with a long history in the world of pharmaceuticals, Dr. Stan Bernard. We not only ask him about the past history of the MSL world, but we also explore his ideas on what the future may hold for the profession at the end of this book.

The Past and Future of MSLs: Interview with Stan Bernard, MD, MBA

Stan Bernard, MD, MBA is President of Bernard Associates (www. BernardAssociatesLLC.com), a pharmaceutical and health care industry management consulting firm offering strategic planning, marketing, medical marketing, competitive simulations/planning, and business development services. Dr. Bernard is nationally-recognized as a consultant, speaker, and author. He has published over 50 book chapters and articles on pharmaceutical and health care topics. Dr. Bernard is a former Senior Fellow at The Wharton School of Business where he initiated and taught in the "Pharmaceutical Management" course for fourteen years.

Previously, Dr. Bernard served as a Consulting Principal at A.T. Kearney and held several executive positions at Bristol-Myers Squibb Pharmaceutical Company. He served as U.S. Product Manager for the launch of the cholesterol-lowering drug Pravachol®, Bristol-Myers Squibb's most successful pharmaceutical product. He served as Associate Medical Director, where he co-founded the first doctor-only Medical Science Liaison (MSL) group in the U.S.. Dr. Bernard also served as U.S. Managed Care Medical Director, the first person to hold such a position in the pharmaceutical industry, and as U.S. Director-Pharmacoeconomics. He also worked in Worldwide Business Development and U.S. Medical Operations. Dr. Bernard received his M.B.A. in marketing and health care management from the Wharton School of Business. He received his Medical Degree from Baylor College of Medicine. Dr. Bernard can be reached via email at: SBernardMD@BernardAssociatesLLC.com and via phone at: (908) 234-2704.

History

As you know, Upjohn started the MSL function in the late 1960's. However, the first MSLs came from sales. Subsequently, other programs moved to professionals with a science background.[2] Can you share your experience on what generated that switch, and how the scientist MSL began?

In 1988, Jan Leschly was the President of E.R. Squibb. A pharmacist by training and an experienced businessman, Jan believed that individuals with combined backgrounds in business and science would represent the next wave of professionals in the pharmaceutical industry and provide Squibb with a competitive advantage. Toward this end, Jan envisioned a sales force composed exclusively of business-oriented doctors.

In 1989, Squibb hired David Best, an MD/MBA from the advertising industry, to turn Jan's vision into reality. David recruited me from Squibb's Worldwide Division to help him create and implement the first **doctor-only**, regionally-based field force. Together we hired, trained, and managed a total of 12 doctors, predominantly physicians. We called this concept "Medical Services Managers" or MSMs to distinguish it from competitors' medical science liaisons ("MSLs"). At that time, MSLs were typically top sales representatives without advanced scientific degrees who called on doctors to communicate scientific information and address more complex product questions.

From the outset, MSMs were designed to be different from MSLs and other sales representatives. As doctors, MSMs could relate to and interact with key opinion leaders and other doctors as "peers". They shared a comparable level of scientific training, experiences, and knowledge with their physician customers. This enabled MSMs to earn the respect and time of physicians. In fact, BMS marketers found that MSMs were spending dramatically more time with physicians than sales representatives.

The roles of MSMs also differed dramatically from those of MSLs. We trained MSMs to handle several novel tasks, such as the management of regional and local opinion leaders, including speaker training; identification and placement of Phase IIIB and IV clinical trials; facilitation of physician dinner meetings; spearheading the pre-launch of new products; and training of sales representatives on disease states and product information. Like a special military force, the roles of MSMs expanded over time as senior Bristol-Myers Squibb ("BMS") executives recognized their extensive capabilities and impact.

It was not long before competitors took notice and began to develop their own versions of the MSM program. The MSM program became the foundation for a number of regionally-based, scientifically-trained medical professional programs across the industry. Probably the most satisfying aspect for me is the number of "MSM Alumni" who have 'graduated' from the program and moved on to executive leadership roles in other companies. While the MSM name has changed back to MSL, the concept of MSLs with scientific training and advanced degrees endures.

Has the MSL changed in terms of work since the first science based MSL teams vs. today?

I think there have been dramatic changes in the roles of science-based MSLs reflecting industry-wide changes. The most significant changes have been regulatory: stricter regulations on pharmaceutical product promotion have virtually eliminated the MSL placement of clinical trials and grants and the facilitation of dinner meetings, for example. Therefore, these MSL roles have been reduced in order to comply with those regulations.

Conversely, three other industry factors have changed to enhance the roles of the MSL: the decreasing numbers and importance of the sales representatives; the increasing complexity of science; and increasing pharmaceutical competition. The declining role and impact of the sales representative means that the MSL interaction and communications with physicians and other stakeholders are more important. The complexity of science has resulted in a greater need for scientifically-educated MSLs to help translate these new technologies and discoveries. New biologicals and biotechnologies, complicated biochemical pathways, targeted therapies, drug safety, and pharmacogenomics are all critical areas that pharmaceutical stakeholders need to understand.

Competition between companies and among products has intensified over the years. Most major pharmaceutical companies need to have a strong MSL group just to be competitive. There is now greater product competition with more in-class and across-class therapeutic agents, generics, over the counter medications (OTCs), and alternative therapies. Because of increasing limitations on promotions, MSLs are playing a larger role in the overall promotional mix for product marketers.

Dr. Bernard's comments on the future of the MSL are contained in the final chapter of this book.

Part I:
Becoming a MSL

Chapter 1:
What is a MSL?

Outside of those performing the job, very few understand the role, activities, or functions of the Medical Science Liaison. How the MSL role is defined and their activities vary widely from company to company. Generally speaking, a MSL is a field-based scientific professional that works for a pharmaceutical, biotechnology, medical device, or managed care company. The primary responsibility of the MSL is to establish, develop, and foster collaborative relationships between thought leaders and the companies they represent via the advancement of science and medicine. Ideally, companies have MSL teams report through the medical affairs department.

In terms of professional background, most MSLs are pharmacists (60-70%), with the majority having a doctorate of pharmacy degree (PharmD). Approximately 10% are PhDs, and 1-2% are MDs or DOs. The remainder of MSLs consists of nurses, or professionals with some type of medical, scientific, or business backgrounds.[3]

MSLs interact with a variety of healthcare professionals. At many companies, the primary contacts or customers of the MSL are academic thought leaders. Academic thought leaders are typically physicians or other healthcare professionals that perform research, publish, and/or teach within an area of expertise. National and international thought leaders often collaborate outside their own home organization in areas of research, publishing, and education. Another customer or contact of the MSL may be the academic institution itself. Many institutions host continuing medical education conferences and need resources for hosting such events. The MSL can assist the institution in finding support for such academic endeavors.

In addition, the MSL may call upon other professionals such as nonacademic physicians, pharmacists, nurses, other allied healthcare professionals or medical education providers. MSLs may also support patient advocacy associations. Examples of such groups include: The American Heart Association, the Cystic Fibrosis Foundation, or The Arthritis Foundation. Many patient advocacy groups also provide educational tools for patients and healthcare professionals regarding the disease states. Finally, MSLs may work with managed care organizations to provide clinical trial information and data for formulary decisions when requested by medical decision makers within the managed care organization.

One of the most common questions asked of the MSL is how they differ from a pharmaceutical sales representative. A sales representative provides samples to physicians. Representatives also sell a product based upon integration of key brand messages and appropriate use of a product based upon the package labeling. Typically representatives work primarily with doctors who provide direct patient care. The sales representative's goals are often based upon their personal and regional sales performance of a drug or portfolio of drugs that they promote. The representative is limited to information included in the approved and current package insert (PI, also known as the product label).

In contrast, the MSL provides medical information, fair balanced scientific exchange, and discusses disease states and all therapeutic options. Since MSLs are scientists, the exchange of data and information is peer-to-peer education and information sharing, rather than selling or marketing. The MSL works within scientific education and research realms. While the primary customer of a sales representative is a clinically based physician, the MSL also works with scientists that perform research and teach. MSLs may also work heavily with pre-clinical researchers, who are in some cases are PhDs. The MSL's performance and goals are usually based upon outcomes, often involving long-range objectives. MSLs do not provide samples, are rarely measured by call frequency and never measured on sales data.

There are 5 basic elements to the MSL role. Each of the 5 elements varies from company to company and between therapeutic areas. The five elements are: research, scholarship (learning), education (teaching), service, and networking (see Figure 1). The five elements are very similar to academic roles and responsibilities. Due to the fact that the roles of MSLs and academics are similar, many MSLs come from academia.

Research. One of the primary functions of a MSL is to foster and champion research ideas and proposals through a company. The MSL literally acts as a 'liaison' between the pharmaceutical company they are employed by and the academic institutions or researcher. Innovation in any field requires constant research and development. Companies seek opportunities to improve the lives of patients and expand product uses through research. Research opportunities are varied and include: investigator-initiated research (prospective or retrospective), identification of investigators interested in participating in company sponsored research, or linking of potential researchers with a shared interest.

Investigator-initiated research (IIR) can be proposed in several ways. Data sets already accumulated preclinically or clinically can be reviewed for new research ideas, or what more commonly is known as retrospective analysis. This data has already been collected either in cells, animals, healthy human volunteers, or patients, and researchers propose to go back and review the data to answer new research questions or do subset analysis. IIR may also

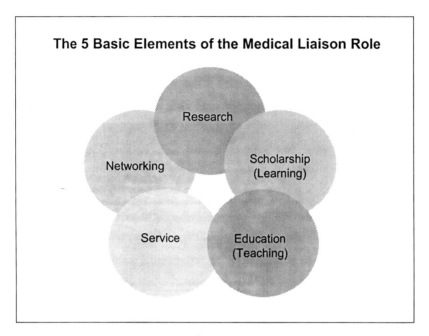

FIGURE 1.

consist of prospective preclinical or clinical trials where the research is yet to be performed. Some researchers focus on preclinical aspects of therapeutic research, others focus on clinical or patient focused research, and some perform both types of research.

Company sponsored research protocols often require identification of investigators. This is another function of the MSL at many companies. Typically, companies sponsor phase I-IV trials for compounds, and need external investigators to provide patients and expertise for clinical trials. Some clinicians only work in clinical research. Other physicians have private practices or university based practices and also perform research on top of their clinical practices. It is the responsibility of the MSL to work in tandem with the clinical research departments to fully understand what the company's needs are for investigators and match those to the physicians performing research within their geographic assignment or territory. Many clinical research departments put together questionnaires or site visit feasibility forms for the MSL to utilize. The MSL may either visit the site to obtain the information, or the MSL may provide the feasibility forms directly to the sites they feel might meet the criteria for a particular study.

Many MSLs have conducted research in the past, particularly if they came from a previous industry or academic setting. In many cases, MSLs have been trained to understand good clinical practices (GCPs) as well as quality research design and proper statistical analysis of clinical research data. Some companies will allow MSLs to co-author papers on original research, and allow for collaborative research and development for all parties involved in a research concept. While co-authorship is somewhat controversial, the point that collaborative research and exploration of new ideas and concepts for improving the lives of patients remains the primary focus and driver for all that the MSL does within the domain of research.

In an industry wide survey of MSLs, investigator-initiated trial work was the most rewarding facet of the MSL role in terms of personal job satisfaction, over other areas of responsibility for the MSL.[3] The intrigue of original ideas and study of research concepts is one of the most rewarding aspects of the MSL role. The amount of time a MSL performs in the area of research will be dependent upon the company, the MSL role itself, and the product lifecycle(s). Newer products tend to have more research questions remaining, which would lend to more investigator-initiated research or Phase IIIB/Phase IV company sponsored research. A product on the market for several years usually has fewer remaining research questions.

Scholarship (Learning). One of the best facets of the MSL job is on the job education. MSLs must keep abreast of cutting edge research and innovation within the therapeutic areas they cover. Therefore, it is imperative that MSLs are life long learners. For those that love education and school, the MSL role is excellent for learning about the latest innovations in science. Intellectual challenge is one of the key drivers for the MSL to have high job satisfaction.[3]

One of the central responsibilities for MSLs is to attend medical meetings and symposia on a local, regional, national, and international level in order to learn about new data and hear what researchers are developing. One therapeutic area of coverage alone could encompass several hundred medical and professional meetings in a year. Since the MSL has technical acumen, they are often the reporters of information from these medical meetings back into the companies, whether it is explicit scientific data gathered at professional meetings, or competitive intelligence gathered in the field. MSLs often attend major medical meetings around their respective therapeutic areas (for example, a MSL in the cardiovascular therapeutic area may attend the American College of Cardiology's Annual Meeting). A listing of major medical meetings and organizations for different therapeutic areas is provided in Appendix A.

There are several industry wide MSL specific meetings that are not therapeutic area specific (Table 1). The majority of MSLs are pharmacists. Therefore, a table is included of major pharmacy meetings of interest to the MSL (Table 2).

Several of the professional organizations listed also have special interest groups or smaller subsets of medical affairs or MSL groups within the larger organization. Those include the American College of Clinical Pharmacy (ACCP), which has an industry special interest group/listserv. DIA has a special interest group for medical communication professionals (the Medical Communications Special Interest Area Communities or SIACs). The Healthcare Business Women's Association contains local or regional chapters for women in specific geographical locations, including Boston, Indiana, Research Triangle Park, NJ/NY, and others. Many of these organizations have not only large annual meetings, but also they host many smaller special interest or smaller geographic meetings, and can be great way to network within the industry.

Education (Teaching). One of the biggest opportunities for MSLs is the ability to share their technical knowledge. Peer-to-peer sharing of information

Industry Wide MSL Meetings		
Company	**When Meetings are Held**	**Website**
Exl Pharma	Fall MSL Best Practices Meeting	www.exlpharma.com
MSL Institute	Several training & MSL manager meetings	www.mslinstitute.com
Center for Business Intelligence	1 or 2 MSL meetings per year	www.cbinet.com
The Drug Information Association	Medical Communications Spring Workshop	www.diahome.org
Pharmaceutical Education Associates	1 MSL meeting in the past - typically fall	www.pharmedassociates.com
Scientific Advantage	Typically have MSL meeting and MSL management meetings in spring	www.scientificadvantage.com
IQPC	Promotion and Liability Issues for Prescription drugs marketed to healthcare conference	www.iqpc.com

TABLE 1

Pharmacy & Professional Organizations

Name of Organization	Acronym	Website	About*
The American Pharmaceutical Association	APhA	www.aphanet.org	"The Association is a leader in providing professional information and education for pharmacists and an advocate for improved health of the American public through the provision of comprehensive pharmaceutical care."
The American College of Clinical Pharmacy	ACCP	www.accp.com	"The American College of Clinical Pharmacy (ACCP) is a professional and scientific society that provides leadership, education, advocacy, and resources enabling clinical pharmacists to achieve excellence in practice and research."

The Drug Information Association	DIA	www.diahome.org	"DIA is a professional association of approximately 20,000 members worldwide who are involved in the discovery, development, regulation, surveillance, or marketing of pharmaceuticals or related products."
Healthcare Business Women's Association	HBA	www.hbanet.org	"Today, the HBA serves its members in a variety of posts and diverse areas, but remains true to its roots: firmly committed to helping women in healthcare advance their careers. Indeed, the HBA's tag line — "Required Experience for Healthy Careers" — clearly articulates the value the association continually delivers to its members."

TABLE 2

* Information is quoted from the organization's respective website

is the foundation of education for the MSL, and is one of the major facets of the role. Many MSLs have come from an academic position (most likely a college of pharmacy), but many come from varied educational roles within and outside the pharmaceutical industry. After a clinical or academic pharmacy position, many move to medical information groups at pharmaceutical companies, and then into the MSL role. Regardless of formal education, most MSLs share a thirst for sharing knowledge and information.

The educational component is somewhat dependent upon product lifecycle as well. For example, a MSL working on a new product may be doing more formal education, such as presentations to managed care formulary experts, or clinicians that are not familiar with the clinical research behind a new compound. A MSL covering an older product may focus less on the basic common information on a compound (absorption, distribution, metabolism and excretion or ADME, pharmacokinetics or PK, drug-drug interactions, etc.) and focus more on preclinical or clinical research that explores possible line extensions of a compound or new indications. While the type of education may change with the product lifecycle, the fact remains that MSLs are educators.

The MSL can also educate internal colleagues at the companies they serve. MSLs are savvy and know what the positions are from many thought leaders on a particular disease state or therapeutic area and can educate on the state of the minds of these thought leaders regarding therapeutics back to the company. This in turn can assist directions for research ideas and collaborations among researchers, clinicians, and pharmaceutical companies. MSLs also attend various major medical meetings (i.e., American Psychiatric Association, The North American Menopause Society Annual Meeting) and can report back to their respective organizations competitive intelligence on potential or competitive products, or provide overviews of the conferences in general to those in the company that were unable to attend the meetings.

Service. Another major component of the MSL's work is that of service. The MSL serves many customers. The first customer of the MSL is the academic thought leader. However, there are many other customers of the MSL, including, but not limited to: internal customers (medical affairs, clinical operations, marketing, sales, medical information, competitive intelligence, drug discovery, pharmacovigilance, epidemiology, etc.), external customers (medical organizations and advocacy groups, professional societies, educators, allied healthcare professionals, and managed care organizations). One of the key components to the MSL functioning well is to fully understand the spectrum of their customers, who they serve, and how to prioritize requests. Whether the MSL is providing a customer with medical or clinical information, collaborating on research ideas or concepts, or sharing external information, the MSL always has challenges in managing multiple customers with multiple projects. The customer mix, both internally and externally, will depend upon several factors: the product's lifecycle, the therapeutic area(s), the size and structure of the company the MSL works for, and the structure of organizations and academic centers within the MSL's geographical assignment.

Networking. Whether or not the MSL is fully aware of it, networking is also a key component to the MSL job. MSLs are connectors. They connect themselves to others, both internally and externally, and link others together where appropriate. They connect to thought leaders, to ancillary staff at institutions, to professional societies, and to patient advocacy groups. They connect the company to various customers in their geographic assignment. They connect thought leaders to each other, sharing research concepts and potential collaborations between researchers, in ways that perhaps even the researchers never thought of before. In many academic institutions, pre-clinical research and clinical research can be disconnected, which is an opportunity for the MSL to cross pollinate, share ideas, and link up potential collaborators. In essence, the MSL is a living, breathing professional networker. MSLs also must connect ideas. One of the skills of an advanced MSL is the ability to think across therapeutic areas, research concepts, researchers, and the companies they represent to come up with scientific innovations and methods of connecting professionals to each other. Referral networking can provide an element of job satisfaction and keep the job interesting.

The Challenges of a MSL

The MSL job, just like any other, is not always easy and perfect. There are challenges, frustrations, and issues with the role. In this section, a few of the big issues surrounding the MSL will be discussed and a few tips and tricks on how to handle these issues will be provided. Other chapters within this book will help with other frustrating areas for the MSL as well.

Damage Control. There is nothing worse than starting out with a new company as a MSL and having the former MSL sabotage your territory before you even step into it. Almost as frustrating is working with a thought leader to submit a protocol idea as an investigator initiated trial or IIT only to have it shot down, or the company has changed its mind during the development of the idea. However, this happens, and it is best managed by damage control. This consists of apologizing for what happened in the past (as it was out of your control), committing to trying to ensure it doesn't happen again in the future, and under promising and over delivering. The cardinal rule of a MSL is to commit to nothing except exploration of possibilities and commit to only what you know **you** can deliver. Promise nothing beyond this and you will be able to successfully manage your customers' expectations.

Metrics. The bane of the MSLs' existence can sometimes be metrics. What the MSLs do is form long-term relationships between the company and the thought leader community. How does a MSL manager quantify the quality of these interactions? How can you benchmark good versus bad work from a MSL? Often, some companies apply the reach and frequency model to their MSL teams hoping that quantity rather than quality metrics are a clear cut, decisive measurement of MSL effectiveness. However, companies that place firm numbers of calls on their MSLs oftentimes alienate the MSL and waste the time of the thought leaders, and run the risk of lowered quality. A combination of quality and quantity metrics can be utilized. MSLs work in a gray zone. The manager that understands this and can in turn articulate the long term value of a MSL team to higher management of an organization will be able to keep their teams in place. If a MSL manager doesn't fully realize the qualitative value of their MSLs teams, the team is doomed before it starts.

Inefficiencies. This and the following section go hand in hand. The adage, "hurry up and wait" applies to the MSL role. As a MSL, there are times that a thought leader will cancel on you last minute, or keep you waiting for hours. The best way to manage this time issue is to make the most of your down time. Make sure you always have work to do. Log your calls or write up your reports while you are waiting rather than postponing it until the end of the week to write your notes. If you delay, you also run the risk of forgetting half of the content of the conversation. Be prepared to have a plan B, and even a plan C. Try to ensure you are going to a place where you have more than one appointment scheduled, so your entire day is not lost if one person cancels.

Planes, Trains, and Automobiles. MSLs travel, frequently and inconsistently. Travel is not what it used to be. Airlines are notoriously late and frankly do not always care whether or not you make your appointment or connecting flight on time. A MSL will miss flights, or spend countless hours (sometimes overnight) in airports waiting for flights to get home or get to the next appointment. You will also travel long distances only to have your appointment cancelled. When you are looking at a potential MSL position, human resources (HR) should be questioned about their travel policies. What type of travel help desk/support do they have in place for the MSL? Do you receive a company car or a stipend for your mileage? Does your potential company offer the MSL a flight club room membership? (Which can be a lifesaver when the airline cancels your flight and you need a quiet respite to complete your work.) If you are happier to have a territory where you drive most of the time, consider items like global positioning systems (GPS), satellite radio, a cell phone car charger, and books on CD.

One of your most valuable items when you are a MSL is your library card. Keep a collection of books on CD in your car ready for a 3-hour one-way trip before appointments. If you can blend learning with travel, you maximize your efficiency while on the road. Many companies offer training sessions on CD for this purpose. Also, if you can keep your frequent user points with one or two airlines and hotel chains rather than splitting them among all the providers, you will have more clout with one (and more points) over the long haul.

Remote (lack of) Control. One of the biggest concerns often heard from MSLs is the loneliness of the job. New MSLs often stepped from a social office environment into the work-alone-from-home environment. Some people never get used to this alienated feeling and discontinue their work as a MSL because of the solitude of the job. The best remedy for this problem is to pick up the phone and call your colleagues. Seek mentors. Find someone within the group that has a lot of experience and ask for help or ideas when you need them. Everyone is better at something. What can you learn from each and every person you encounter? This also includes your thought leaders. Like other professionals, thought leaders are the best at what they do in a certain area and they like to teach, so ask a lot of questions and learn from them.

Paperwork. The pharmaceutical industry is one of the most regulated industries in the world. PhD bench scientists keep scientific notebooks. PharmDs have to keep records of prescriptions being filled. MDs have charts for their patients. Documentation for the MSL is a must; there is no way around it. Manage it. Fill out your forms and reports as quickly as possible. Find a best way for you to stay organized and be regular about your reporting. It is important to record expenses after each trip. Submit expense reports with frequency and regularity. Get into the habit of doing paperwork in a timely fashion. Your manager will thank you, and you will thank yourself in the long run. Also, it is important to schedule regular, reoccurring times during the week to be in the office. Office time is just as important as field time for the MSL. A MSL that is in the field 5 days a week will struggle to find time to file their paperwork and keep up with their reading, unless they enjoy working on weekends, which can lead to burnout.

Job Security. MSLs, in general, feel that their jobs are not secure. In our annual MSL job satisfaction survey, job security was one of the higher areas of concern to MSLs versus the general population utilizing a model of Hertzberg's theory of job satisfaction.[3,4] With globalization, unanticipated Food & Drug Administration (FDA) rulings, and the unpredictability of research outcomes, the one thing a MSL can count on in this job is change. The best way to manage this issue is to be flexible, become the best at what you do, and not fear or resist change. Nothing is guaranteed in life, but if you can make yourself an expert in a specific area of the MSL role, or if you can go above and beyond your job description, smart managers and smart companies will bend over backward to keep you.

Change Management. As mentioned in the last paragraph, change is inevitable as a MSL. The industry is dynamic. If you like science, business, and change, the MSL role could be a great one for you. Ask yourself what new trends you are interested in learning more about and how they can fit into your current role. Seek better or more efficient ways to improve your work. Try new ideas and concepts out and share them with your colleagues. The best way to manage change is to embrace it and be as flexible as possible.

Moving on Up, or Over. The most frustrating item for a seasoned MSL is where to go from the MSL role. Many MSLs feel they are caught in a "career cul-de-sac" and can't get out of it. Perks like working from home, a company car, and autonomy are very seductive. If you feel you are stuck in this situation, realize that every job has plusses and minuses. Can you improve your current situation by taking on extra projects? Can you further your professional coursework? Are you fully utilizing the tuition reimbursement supplied by your employer? Ask yourself what motivates you. Is it working from home? If so, then find a different type of job that you can do from home and enjoy. If you thrive on intellectual challenge in your work, take a class. If it is the travel you enjoy with your MSL job, can you travel and enhance your work at the same time? If you like the autonomy of the job, have you thought about starting your own company? Even though your title may stay the same, as a MSL, you can work on many internal projects while keeping your traditional work flowing. What type of internal work will make you engaged and interested? Some MSLs have made their entire careers out of being a MSL, and there is absolutely nothing wrong with being a career MSL. The challenge is finding out what makes you personally tick and maximizing that part or facet of your work while minimizing or eliminating the work you don't enjoy performing.

The Real World: An Interview with a Working MSL - Ernie Pitting, RPh, MS, PharmD candidate

Ernie is currently a MSL covering the central United States for a small pharmaceutical company. He began as a registered pharmacist working in various positions, including decentralized hospital pharmacy, community pharmacy, and various roles within industry. He has worked in market research, and spent many years working his way up through the ranks of the pharmaceutical sales realm, both in general and specialty sales. He holds a BS in pharmacy, and a Master's in Science from the University of Wisconsin in Pharmaceutical Marketing, and is currently working on his PharmD.

How long have you been a MSL?

Five years. I have observed many colleagues move on to different companies. In conversations with my former colleagues I have discovered each pharmaceutical company utilizes their MSLs in very different ways. In fact, even their titles within each company differ.

You came from a sales position. How has sales helped you become a MSL?

There are multiple aspects that are helpful. It is beneficial because: 1. Being a representative helps you understand the foundation and basic work of the company and the industry (sales, marketing, research and development, medical affairs, etc.). 2. More specifically, it helped me to understand how to make presentations one on one in a professional, yet conversational manner and how to tailor that presentation as you move through it, based on your reading of your audience of one. 3. It taught me how to be comfortable in front of physicians and gain access to them, and not think of them as MDeities. Some pharmacists are unfortunately not comfortable in front of physicians. 4. I have come to appreciate that the landscape of knowledge and interest is vastly different between community and academic physician practices. 5. I gained an invaluable working knowledge of the important interplay between field-based staff and home based staff (as I say, "in house and out house people"), and how to get things done. For example, how to gain agreement between various factions to get a pilot study done. How to build a consortium of financial support to secure funding for an important educational endeavor, as another example. It is very important to know how companies in general and how your specific company gets things done, and how it serves its physician and patient customers.

While MSLs are definitely scientific and fair balanced in their focus, we do so within the workings for a proprietary company. As such, I think in the MSL role, it is important to understand the role of pharmaceutical representatives. That doesn't mean you become a salesperson as a MSL. However, you can empathize and work with them because you understand their needs better. While the roles that MSLs and sales representatives execute are quite separate, they are each best served by engaging in some degree of collaboration.

What are your professional rewards from the job?

I am rewarded by my collegial relationships with high level thought leaders in different therapeutic areas. I have the luxury of knowing a little about a lot and talking to thought leaders at a high level. That is cool for me. I also love connecting the dots - from management of patients, to thinking about different therapeutic areas, to pilot studies that may progress to bigger National Institutes of Health (NIH) funded studies down the road. It is fun to have conversations with physicians about data. I have a great passion for the disease states I work in and enjoy my conversations. The firm I have been with has helped me develop a passion for my therapeutic area. The face recognition I have with thought leaders outside of my immediate geography is rewarding as well. I am on a first name basis with many of these national experts. That is really cool and one of the rewards of my work.

Tell us about your educational background and how it has helped you as a MSL.

My perception is that the standard for MSLs will be a doctorate degree. My newfound interest in my PharmD program is that nearly every therapeutic area I've reviewed for school, I have found some pearl or interesting piece of information that I've been able to put into my own therapeutic area as a MSL. It has helped me become better at what I do - I am a better MSL through this educational process. I can talk at an even higher level in basic and clinical science. For example, I was at Mayo talking to a group in a scientific arena that was cutting edge. Their more tangential issues are now quite understandable. As such, being that drug information expert with my thought leaders has been great. I'm getting more current information than some of my colleagues who obviously went to school several years ago. This too is very helpful.

What do you like best and least about the MSL job?

The best part is professional fulfillment as I mentioned. It has given me a better pay scale than I've never had. I am frankly paid well for work I love to do. I have found where I'm supposed to be in life!

The downside of the job is the world that I work in is always 'gray,' unlike true pure research and development, or pure marketing. It can be frustrating at times as we are pulled and pushed in different directions. It is important to always remember that you are hired because you are a professional, and as such, are paid to make those judgment calls based on your professional background. I'm now covering 13 states, so the geography has been a challenge. It is really cool to go to some other major medical centers, but travel isn't as glamorous as it used to be. Another challenge is that some clinicians don't value MSLs. Not all clinicians find the same value in the data or in information you can share. Corporate politics are a challenge at times as well. Achieving a work/life/school balance has also been a challenge. For example, working at home can become obsessive; there is always something else to be done. Medical conferences are on weekends. The impact of all this travel can have negative implications for significant others. Turnover and change is challenging for others, whereas I have not yet had this as an issue.

What advice would you give to a pharmacist interested in this role?

Advise them that working for industry is not a bad thing. We all work for money, but working for a company can stimulate science. Perspective is really important; everyone is selling something! The challenge of working as a MSL is that you have to be really self motivated. If you need someone to help you show up on time or if you need to work in a more personal, every day community-ish setting versus a virtual community; this may not be the role for you. If you are a self-starter and like freedom and variety and can manage the balance of travel and life, as well as work in a more virtual working environment, this can be a marvelous position.

It takes at least an entire year before the MSL job clicks. You need to remain in a job for at least one year to figure out whether or not you like it. It takes awhile to get it all together, so you do need to give it a chance. You have to be willing to motivate yourself enough to keep learning. To not know the answer to something once is acceptable, but not twice. This isn't emergency medicine, but one does need to be self-motivated in order to perfect the work and the craft of being a MSL.

Take advantage of all the resources you can to expand your knowledge. I don't think we had the opportunity to pull apart studies in pharmacy school. We also didn't get enough training in developing a question for research, putting that protocol together and then doing the work. After that there is writing of an abstract, getting it accepted at a meeting, developing a poster, and finally writing the manuscript. It is very important to understand the world of published medical science. The world of medicine is driven by data - to understand and value the literature is key. Using textbooks and professors does NOT mean one is reading the primary literature.

In the current environment, it is interesting to appreciate the evolving world of medicine. Academicians have become even more strapped for time due to training and clinical responsibilities, as well as traveling and speaking. They simply have even less time to read the literature. This has become a huge opportunity for MSLs as they can provide updates for the academicians. I see this as an ongoing trend. Even national thought leaders don't have the time to read their journals.

The other challenge is technology related. Technology is everywhere. It has become a little too easy for people to communicate. As such, managing the multiple streams of communication is becoming a challenge for all of us. In addition, everyone expects immediate access, and unfortunately that's not always possible when working in the field.

You mentioned the variety of work within the scope of the title "Medical Science Liaison."

Yes, I think it's changing and evolving. Interview with many companies and ask if you can talk with a MSL currently on the team to describe the job. Every company does the job differently. Some companies have their MSLs focus more on research, or more education, or more presentations. Some companies have their MSLs do all of these functions. I have also come to appreciate having "tough skin" in this work. Don't take comments too personally. Working in sales helped me immensely with this issue. Sometime someone will take his or her bad day out on you. It is important that you remain confident in what you are doing--one must stay resilient.

I think it's a really great role for someone that is independent. If you like knowing a little about a lot, if you like working with professionals in different arenas (sales, marketing, academia) and don't mind the travel or the lack of an office family, it's a cool role. The industry is really not the devil incarnate. It is important to have perspective on the values that the industry brings at large.

When I was a sales representative many years ago, I would spend 20 to 30 minutes discussing a new article from a journal with a physician. However, the world of medicine has drastically changed since then. Sales representatives may get only get 10 seconds of sound bite selling. So now the old representative's role is the new MSLs role. Many years ago, Upjohn and Lilly employed only pharmacists as sales representatives. Now it's been elevated to the MSL role.

Just another point, physicians don't like turnover. They do like to develop relationships with colleagues in industry. In addition, many doctors love to see PharmD's who understand clinical practice. They want clinical experience and relevant patient practice conversations based upon the literature. Simply put, I think this is a fascinating role for pharmacists out there that wish to challenge themselves with work that is not black or white.

Chapter 2:
Activities, Salary, Job Satisfaction and Career Paths of the MSL

MSLs can have a variety of professional backgrounds. In this chapter, data will be shared from the author's 5-year longitudinal MSL industry wide job satisfaction survey, as well as the features and benefits of the MSL role. A listing of several actual career paths that various MSLs have taken is provided in Appendix B of this book, to illustrate the variety of candidates for the MSL role. Profiles of the following professionals: MD, PhD, PharmD, BS Pharmacy, DVM, RN, and other non-medical professionals are included.

The data presented in this chapter is a culmination of 5 years of industry wide surveys regarding field based medical science liaisons between the years of 2003-2007.[3,5,6,7,8] An online survey website was used to collect the data (either SurveyMonkey.com or Zoomerang.com, depending upon the year). During the last two years of the survey, participants were not paid, but were offered a choice of charities to fund for their time in filling out the survey. Sample size ranged from 107 participants the first year (2003) to a peak of 141 participants in 2006. Each year had over 100 participants.[3,5,6,7,8]

Demographically, there is a range of professional backgrounds for medical science liaisons. In the 2007 survey, 65% of responders held PharmDs (N=76), followed by BS Pharmacists (20.5% N=24), PhDs (15.4%, N=18), MS degree holders (12% N=14), MBAs (10.3%, N=12), and 'other' (8.5% N=10). Other backgrounds with responses less than 5% included: MD, NP, RN, BSN, and MPH. The 8.5% that chose 'other' described their degrees or backgrounds as: MSN, MS, MA, BA, RD, BS, CDE, ASRT and fellowships. Percentages did not add up to 100% in the survey, as multiple responses were allowed. Eighty-two percent of responders in the fifth year survey responded with a doctorate degree (PharmD, PhD, or MD). [3,5,6,7,8]

One of the first evolutionary requirements for MSLs was an entry-level doctorate degree, but a possible second evolutionary step for MSLs may be board certification. Twenty percent of the responders to the 2007 survey were board certified.[8]

Most frequent responses were BCPS and BCPP for pharmacists in this survey. Currently, there are 5 recognized specialties for pharmacists, according to the Board of Pharmaceutical Specialties (www.bpsweb.org) and they include: nuclear pharmacy (BCNP, established in 1978), nutrition support pharmacy (BCNSP, established in 1988), oncology pharmacy (BCOP established in 1996), pharmacotherapy (BCPS established in 1988) and psychiatric pharmacy (BCPP established in 1992). Pharmacists can also be certified in toxicology [Diplomat of the American Board of Applied Toxicology or (DABAT); more can be found on their website, www.clintox.org]. Pharmacists can also be certified diabetes educators (CDE). The website: www.ncbde.org has more information on this process. Pharmacists can also have post-doctoral fellowships. Nurses and physicians can also be board certified.

Of interest to those considering the MSL role is the fundamental question asked all five years: How satisfied are you as a MSL with your current position? Generally, the MSLs are a satisfied lot. Over the surveyed 5 years, the majority of MSLs have either been very or somewhat satisfied with their work (81-89%). Only 1-5% were very unsatisfied in the survey. Length of tenure was also asked, and most in the fifth year's survey were in their current MSL roles for 36-60 months (nearly 20%), which has been consistent over the entire 5 year survey length. When asked about area of work prior to the MSL role, the most frequent answer was nonindustry, clinical pharmacy (2003=14%, 2004=25%, 2005=17%, 2006=not asked, 2007=24%). The most frequent response after clinical pharmacy was academic pharmacy (19%, 19%, 20%, na, 14%). Other common pathways from industry included: medical information, clinical operations/research, marketing and sales, and pharmacology/toxicology. Nonindustry routes included private practice and managed care. In 2005 and 2007, respondents were allowed to choose 'other' and explain what other included. Responders wrote: medical writing, health care web site, hospital, consulting, fellowship training, long term care consulting, medical education, industry managed care, dot com company for electronic prescriptions, and sales. The good news is that there is a lot of variety when it comes to professional backgrounds of the MSL pool.[3,5,6,7,8]

The typical response rate for length of time a MSL is planning on remaining within their current MSL role is greater than 5 years (32.5%, N=38, 2007 survey). A quarter responded they would only remain in their current MSL role 1-2 more years. When asked about whether or not the MSL's current company has a formalized career path for MSL advancement, half responded in the 2007 survey the company did not, while 15% responded they have a formal path both inside and outside of the MSL function, and the remaining 35% responded there was only a formal career path within the MSL related function.[8] This is an area of opportunity for pharmaceutical companies to assist in development of career paths for talent from the MSL organization. Companies that have more established pathways can potentially retain MSLs over the long term because MSLs have more clearly defined pathways for career advancement. MSLs oftentimes feel they are stuck in a "career cul-de-sac" and have limited career advancement outside of the MSL role.

When asked about what position the MSL would like as a next step in their careers, most responded they desired a senior MSL position (32%, 2007 survey). Sixteen percent desired a MSL manager role, followed by 'I don't know (12%), management in pharmaceutical industry (7%), clinical operations/research (6%), and consulting (5%). Other responses under 5% included such work as: in house medical affairs position, training, regulatory affairs, academia, managed care, product development/business development, retirement, medical advisor, DEA, entrepreneurship, or health outcomes.[8] There is an entire chapter later in the book devoted specifically to next steps for the MSL (Chapter 10).

Many MSLs move from one MSL position to another within their career. Survey results show that 23% (2003), 22% (2004), 39% (2005), 30% (2006) and 41% (2007) of MSLs have worked for more than one company as a MSL. The most common reason why MSLs left their previous MSL posting in the 2007 survey was a company merger, reorganization or downsizing (45-47%). Other reasons in 2007 for leaving included: compensation (23%), corporate culture (20%), lack of pipeline/new data/product in lifecycle (19%), lack of intellectual challenge (17%), unclear direction on future positions (17%), manager (17%), poor or lack of a bonus (17%). Other less frequent reasons included: burnout, too much travel, too many hours, poor/lack of stock options, lack of trust, lack of empowerment, revision of the MSL role, retirement, glorified sales position and 'other'. (Other included: promotion, concerns about company, large geography, and a better opportunity with another company.)[3,5,6,7,8]

Size of MSL organizations has held consistent across the 5 years of surveys. When asked how many MSLs in your company cover the US in your group, the most frequent response in 2003 was 30 or more (31%), (29% in 2004, 30% in 2005, and 40% in 2006). In 2007, the most frequent answer was 6-10 MSLs per group (26%), followed by 30 or more (24%). Although the sample sizes are relatively small, possible explanations for the changes in size of groups between 2006 and 2007 could include smaller groups within a company having bigger geographies, or older products being supported by smaller forces. When asked how many MSLs does your company employ? twenty seven percent responded 51-100 MSLs, and 27% responded 101-200 MSLs in 2006. Thirteen percent of responders stated their companies employed over 200 MSLs. In 2007, the most frequent response was 1-25 MSLs total were employed by the company (32%), followed by 51-100 MSLs (24%), 101-200 MSLs (17%), and 26-50 MSLs (14%), and over 200 MSLs (14%). A literature search revealed no statistical data regarding the total US or global population of MSLs, but one can guess that it is between 5,000-10,000 MSLs globally based upon anecdotal information gathered from experts in the MSL recruiting field.

In terms of therapeutic areas, most MSLs only cover one therapeutic area, although in the first year (2003) survey, the majority of responders covered two therapeutic areas (46%). Those responders covering one therapeutic area over the life of the surveys were 25% (2003), 56% (2004), 60% (2005), 64% (2006), and 54% (2007). When asked about products covered by the MSL, most responded that they cover two products 46% (2003), 26% (2005), 20% (2006), 34% (2007). When asked about amount of overnight travel per week required by the MSL job, the consistent response was 2 nights per week (31-36% for all 5 years' surveys). [3,5,6,7,8]

A series of questions was posed about customers. When asked whether or not MSLs had a designated opinion leader call panel, 80% responded yes in the first 3 years of the survey. In 2006, sixty percent responded yes. In 2007, 70% percent responded yes. The majority responded they called upon 20-50 thought leaders in all 5 years of the survey (60-75%).[7,8]

Part of the responsibility of the MSL is to cover major medical meetings. In 2004, a question regarding attendance at national/international medical meetings the MSL attended per year was included. Most in 2004 attended only 2 national/international medical meetings per year (31%). In 2005, nearly 1/3 of responders reported they attended more than 4 meetings per year. In 2006, the question was widened to include greater than 7 meetings per year. Fifty percent attended 3-4 meetings per year, while 32% attended 1-2 meetings per year in 2006. In 2007, 39% attended 3-4 meetings per year, while 41% attended 1-2 meetings per year. We also asked about number of hours worked per week in all 5 years. The majority responded they worked between 46 and 50 hours per week.[3,5,6,7,8]

The number of times per year the manager joins the MSL in the field for collaborative work has not been consistent. In 2004, the most common response was 3 times per year for joint calls (27%). In 2005, the most common answer was never (22%). In 2006 and 2007, the most common answer was 4 times per year (18% and 22% respectively). Sixteen percent stated they never worked with their manager in the field in 2006 or in 2007. Obviously, the MSL role is not similar to sales, and neither is there any consistency of MSLs working with management across the industry.[7,8] It is interesting to note that nearly 1/5 of all MSLs responding to the survey have never worked with their managers in the field. While 4 times per year may be extreme, never working with a MSL and manager combination in the field begs the question - how can the MSL be assessed as to their effectiveness?

When assigning a ranking of the MSL job elements to individual job satisfaction, the elements MSLs enjoyed most in the survey included the following (in order of most satisfying to least satisfying):[7,8]

1. Opinion leader relationship development
2. Investigator-initiated trial development
3. Attending medical meetings
4. Advisory board coordination/participation
5. Training others (inside the company)

6. Lecturing/educational programs - non CME (tie)
6. Receiving training (tie)
7. Company sponsored investigator site identification
8. Support of company sponsored trials
9. Participation in scientific meetings/posters/abstracts
10. Lecturing/educational programs - CME (note that 31% marked this not applicable)
11. Training others (outside of the company)
12. Internal department project work
13. Attending internal company meetings
14. Travel
15. Administration

Assessment and measurement of MSL performance is one of the most elusive elements to the work. Many MSLs and MSL managers struggle with the grayness of the role. Geographic differences and therapeutic differences can make one territory great for one MSL in one geography, while for another MSL in a different therapeutic area can have a less interesting assignment. For example, comparing MSLs on investigator-initiated research for a Boston territory (one rich with research opportunities) verses a western territory (one with not as many universities and research centers) can be a challenge. A MSL in California with a lot of managed care presentations may need a managed care background, while a Florida territory may require less managed care experience, but more geriatric experience. It is difficult to demonstrate consistent performance across geographic and therapeutic areas when there are so many variable factors. These caveats being taken into consideration, MSLs were asked how they were currently being assessed or measured in their positions. The most common response in the 2007 survey was by activities & outcomes with opinion leaders (73%), followed by the ability to meet preset objectives (65%), number and frequency of calls (65%), interaction with internal team (60%), projects outside normal scope of job (44%), identification of thought/opinion leaders (43%), ride alongs with management (39%), feedback from thought/opinion leaders (37%), the number of programs given/supported or organized (31%), goals for territory (25%), number of protocols submitted (23%), sales and/or company performance (9%), and other (8%). 'Other' included: company sponsored trial success, number of face to face meetings per quarter/field days per year, quite honestly it's unclear, making slide presentations, feedback from internal colleagues, driving record, and number of abstracts/publications.[8] As an interesting side note, the notion of MSLs co-authoring publications while working for a pharmaceutical company is a controversial one. Some companies believe MSLs should publish or perish, just like academia. Other companies feel their MSLs should not be publishing and authoring original work due to possible bias. In the opinion of the primary author, the MSL should be coming up with and seeking original research and ideas. MSLs truly must have deep technical expertise in whatever therapeutic area(s) they work in. Without original research in their past and current work, how can they be experts within an arena? Even if they write review articles, opinions, or any publications, they are peers in research rather than mere administrators to the research process.

Why are MSLs satisfied with their work? Most common reasons in the 2007 survey included: the ability to work from home (88%), intellectual challenge (78%), enjoyment of working with opinion leaders (77%), learning while on the job (62%), enjoyment of working with peers (60%), great manager (53%), great company (41%), compensation is great (40%), ability to move to other positions within company (20%), great mentor (15%), and other (3%).[8] In previous years, the most frequent answer to this question was intellectual challenge, but it has subsequently been surpassed by the ability to work from home.[7]

If you ask the typical manager, they may already know about Herzberg's model for discussing motivation of employees. If you would like to know more about Herzberg's theory, please refer to his classic Harvard Business Review publication, printed originally in 1968 and reprinted in January 2003. (It can be found online at Harvard Business Review's website, the reprint number is RO301F.)⁴ Herzberg's identified dissatisfiers or hygiene factors (those elements within or around a job which can be changed but probably will not bring about more job satisfaction) as well as his motivating factors (those elements that by being changed can bring about more job satisfaction). This model and results were compared to survey results to see if there were differences. Some differences were identified. In terms of the motivating factors, MSLs most value personal achievement, followed by responsibility, growth, the work itself, recognition, and advancement. In terms of hygiene factors, MSLs are most frustrated by company policy and administration, security, personal life, salary, supervision, status, work conditions, relationship with supervisor, relationship with peers and relationship with subordinates. As the MSL role is field based, the traditional office environment does not apply. The wise MSL manager would focus on keeping the opportunities for achievement, responsibility, and growth maximized, while trying to remove bureaucracy, paperwork, and ensuring some stability in the work for MSLs. As previously demonstrated through the data, most MSLs left their last role not out of choice, but because of corporate mergers, downsizing and/or reorganization. The leading components of current MSL positions that make them least happy in 2007 included: regulatory guidelines (22%), administrative tasks/paperwork (16%), lack of advancement opportunities (10%), and lack of new data or pipeline (7%).[8]

In terms of training, MSLs wanted further training according to the rank below (from most desired to least desired):

1. Disease state/therapeutic information
2. Competitive intelligence/competitor clinical trial information (tie)
2. Opinion leader collaboration development (tie)
3. Table or facilitation of small group skills
4. Clinical trial design
5. Presentation skills

6. Statistics
7. Business plan development
8. Time management/organizational skills
9. Operational training

Other types of training included: pharmacoeconomics/health outcomes, communication skills, identifying and utilizing existing strengths, computer optimization, leadership skills, drug development process, clinical trial operations and regulatory training related to clinical trials (Good Clinical Practices or GCPs), training on pipeline, soft skills like negotiation and listening, and emotional intelligence teambuilding process mapping.[8]

The surveys ended with demographic information regarding MSLs. Most MSLs are not currently seeking other employment, but nearly 50% would consider other opportunities if called by a recruiter. This may mean that although the majority of MSLs are satisfied with their work, job satisfaction does not directly correlate to loyalty. In the free agent environment MSLs work in, the MSL manager would be wise to realize she or he is vulnerable to market forces taking away his or her talent. It is critical for the MSL manager to fully understand what is of importance to each of her or his MSLs individually. One MSL may highly value more vacation time, while another may be motivated by salary, while yet another could be interested only in working on new, creative projects. Better understanding the individual's motivators can provide information for the MSL manager to keep her or his team motivated.

Fifty to sixty percent of all respondents to the survey were female. In 2007 52% were female, and 59% were female in the 2003 survey. Base salary results by gender are presented in Table 3. The disparity between female and male salaries has narrowed; however, female MSLs are still paid less ($0.97 for every $1.00 a male MSL is paid in 2007), and the gap between genders in the MSL field is narrower than other professions. A limitation of our studies is that information regarding number of years' experience was not collected in all five surveys; thus, we cannot compare salary to gender and number of years' experience, which could provide a large variability. [3,5,6,7,8]

| Year | Medical Science Liaison Base Salaries ||||
	Overall average base salary	Male Average	Female Average	Difference
2003	$84,785	$88,140	$82,500	$5,640
2004	$87,900	$90,240	$84,500	$5,740
2005	$99,084	$100,800	$97,600	$3,200
2006	$104,091	$110,000	$98,714	$11,286
2007	$109,348	$110,357	$107,845	$2,512

TABLE 3. Source: *2003-2007 MSL Job Satisfaction Surveys.* [3,5,6,7,8]

Benefits for the MSL: Interview with Chris Conley, CSAM

In this interview with Chris Conley, a recruiting professional, we focus specifically on the benefits of a MSL as they relate to the work and job satisfaction of the MSL.

Tell us about your background.

With 17 years' experience as a proven professional in biopharmaceutical recruiting, I have a vision for excellence, the passion, the talent and the drive to carry it out. I started my career in information technology (IT) contract recruiting and realized I didn't like this industry. I landed a contract with a small biotech company and the rest is history; I found my passion. The majority of my biotech career has been with MRI where I have built several long-term relationships with clients to the point I've become a consultant. I specialize in Medical Affairs, Marketing, Research & Development, and executive level positions. I have produced numerous successful biopharmaceutical divisions and work only with the biopharma's elite. My success stems from taking a consultative approach to recruitment. I take great pride in working with my clients--both candidates and companies.

With my understanding of the biopharmaceutical industry, I will work on retained and exclusive engagement searches specializing in project team building. I am involved in mentorship programs and am recognized as an industry expert in biotech and pharmaceutical recruiting and consulting.

This chapter included motivators for MSL job satisfaction. Money is not everything to the medical information professional, but let's focus on benefits other than direct compensation. What benefits are key for a MSL to consider beyond base salary?

Sign on Bonuses: Not all companies offer sign-on bonuses. Generally speaking, you have to be leaving something on the table or possibly relocating to receive a sign-on bonus. They are becoming rare unless the expansion falls in the 4th quarter then you have more leverage because the annual bonus is at stake.

Stock Options: These come in the forms of grants, options and restricted stock units (RSUs) and depending on the company, it is a way to lure quality MSLs over or a way to incentivize your existing team, a built-in golden parachute if you will. All companies that I work with offer stock options. A general rule of thumb, the smaller company with higher risk offers more stock options.

Annual performance bonuses: Literally every company, pharmaceutical or biotech, offers annual bonus incentives. Big pharma offers lower bonus potential that runs around 12 – 16%, whereas biotech companies typically offer 20 – 35% bonus potential. Some companies offer profit sharing, which is a combination bonus, based on the company performance for the year and the personal performance contribution by the individuals.

Vacation: This varies from company to company. Typically companies offer 27 days per year. There is a combination of holidays that are honored by the company and the standard 2 weeks vacation pay that is accrued. Some companies offer 3 or 4 weeks vacation to their senior level MSLs and sometimes there is a carry-over that is honored, but it is negotiated up front in the recruitment process. Some offer to allow employees the opportunity to purchase additional vacation time.

Medical, dental, vision and mental healthcare: Health care benefits are the gold standard. They are effective immediately and cover the individual. Some companies offer health maintenance organization (HMO), or preferred provider organization (PPO) choices. These benefits cover the employee and often will have a co-pay coverage for the employee's family. Depending on the size of the company, the larger companies are self-insured. Some companies offer adoption assistance, and all offer maternity and paternal leave. Every company that I work with offers tuition reimbursement, short and long term disability, sick leave, and 401K. On that note, some match more on the 401K plan. One company offers retirement benefits after age 55 no matter how long you have been with the company.

Vehicles: Since these companies know the benefit of a MSL, they all either offer car allowances ranging from $700 to $1100 per month, or they offer an upscale car like a Volvo or an up-line sport utility vehicle. All MSLs receive a home office set up: phone, cell phone, fax/copier/printer combination, laptop, and Internet access. Most offer a decent travel policy where if you earn sky miles during work travel, you can travel for personal reasons using your miles.

Other benefits to be considered include:
    Wellness programs/gym discounts
    Child/elder care
    Sabbaticals
    Research funds
    Community/not for profit service and donation matching
    Employee referral plans

How flexible beyond base salary are companies for negotiating other benefits?

It depends on the size of the company. If it means that corporate policy will be affected then they have to stand on policy. Generally, the smaller the company, the more flexible they are. The larger companies are less likely to be flexible due to the size of the company and if policy is changed, then it means everyone in the company is eligible for the same.

When is the right time for the MSL to begin negotiating benefits?

It is always wise to tell your recruiter up front before you are submitted to the position what your expectations or requirements are. Sometimes, candidates have unrealistic expectations and companies cannot meet these expectations. This is more than likely the cause of a bad recruiter spreading untruths about the position. For instance, I'm not sure where this urban legend originated; however, there is no written rule that states when you make a change to another company that they are required to give you an increase by 15%. The candidate's own company doesn't give annual increases that amount to more than 3.5% so why should the new company be expected to take give that size of an increase?

Once your recruiter knows what it will take for you to make a change, they should state up this up front to the employer if it means you will not accept the position if it were offered to you. If you have a trip planned in August and it has been planned for over a year and you will not qualify for vacation time if you start in July, then this should be stated up front. Generally, it is the manager's decision at a business level.

How should the MSL determine what is important to them regarding benefits?

I believe it is a personal issue for each individual. Some have families with special needs that require a certain level of medical insurance coverage. I would not do a side-by-side comparison since each company offers something different. It is hard to compare one company to another. The only time I would actually look hard at a company's insurance benefits is if they were a small company and didn't offer a competitive package. At that time it is appropriate for a sign-on bonus or additional monthly income to keep them whole.

What about relocation and a job offer? Are relocation packages a thing of the past for MSLs?

Most companies offer relocation packages. Each is different. Unfortunately, it is extremely rare to find a company who will pay an executive relocation. Most companies will either give a one lump sum or they will pay for pack and move with some additional expenses. I rarely see companies paying closing costs on either end of the move. I do, however, see companies paying for housing assistance and for one or two trips out to the relocating city. It has been years since I have seen companies pay large down payments on homes for relocating candidates. Some companies will offer mortgage assistance, which is a fixed or lower rate if an employee has been with the company for a number of years. It isn't common and it is typically done by California based biotech companies who need to provide incentives to candidates to move to their state. This is due to the high housing costs.

A MSL's value is in their therapeutic specialty and their established Rolodex™ of thought leaders. It is more likely to see managers or directors relocated. I have relocated candidates in a MSL role to Boston, to Minnesota/Wisconsin, and San Francisco. These are the most challenging territories to find talent in so you are more likely to see relocation in these territories.

Are there any other myths about MSLs or candidates in general that you can debunk?

Since the industry has changed, I understand the need for the terminal degree requirement. That doesn't mean I agree with it, I do believe there are extremely talented individuals who have performed exceptionally without the "D" behind their name. There are few companies who don't stand on this requirement. You run the risk that if your company is acquired the first MSLs cut won't have the terminal degree. If a candidate or current MSL were planning on staying in this kind of role, then I would highly advise getting a terminal degree.

Not all recruiters are bad. In fact good ones are just as important to your career as a doctor is to your health. If you are considering making a change don't send your CV into the black hole of career websites. Work closely with your recruiter. Start early so when you get to the point where you want out you have a new place to go. Your recruiter owes it to you to keep you informed. Do not send your CV to everyone you talk to and PLEASE NEVER put your social security number on your CV. This is the only change I ever make to my candidate's CV. I remove it immediately. With all of the identity theft I do not want that responsibility of having that information. We don't need it so only add that on your application when you are interviewing with the client company.

What other advice do you have for MSLs or potential MSLs?

Network. Meet new MSLs at your conferences. Get to know your competitors; they can be your next employer. The MSL world is a small world. I cannot believe how many MSLs know each other. Besides, you never know when you will be on the hiring end and will need to know these people! This is an amazing career and I enjoy working with each and every one of the MSLs that cross my path. I would like to know exactly how many MSLs there are in industry to date because it is my goal to meet every single one at some point in my career.

Chris can be reached at: 503-317-1901 or via email at: Dna2c@comcast.net.

Chapter 3:
Getting Hired

As a consultant to many budding MSLs, one of the biggest questions Erin often receives is, "How can I become a MSL if no one will hire me unless I have MSL experience?" Many are frustrated as it appears to be a 'catch 22' of only being hired with experience. However, everyone has to start somewhere, so how can one break into the industry at the beginning? Obviously, this hasn't been the case since the inception of the MSL role, otherwise, there would be no MSLs at all. Not everyone needs to be a previous or current MSL in order to obtain work as a MSL. However, the first job is often the most challenging to obtain.

Even before you decide if a MSL role is right for you, a bit of homework is required. This chapter will review the process one must go through in order to not only obtain a MSL job, but more importantly, to better understand if the role is even right for the individual before they begin applying for positions. There are a few steps to this process: 1. Understanding your personal values, 2. Understanding your personal strengths, 3. Writing a cover letter (because without experience, you most likely won't be working with a recruiter), 4. Networking strategies for the MSL role, and 5. Websites and resources helpful to better understanding the world of the MSL.

Part 1: Values Assessment

The role of MSL is a fabulous career choice. It offers a wonderful blend of business acumen, cutting edge medical research, and networking skills. The same day never occurs twice, and the MSL enjoys autonomy, creativity, and intellectual challenge. Before you start looking for a job, be it a MSL job or any other job, stop and ask yourself this question: Do I know what I value? If you do not have an answer, finding a job that is personally and/or professionally rewarding may be treacherous.

The MSL job is not for everyone. Many individuals have started a MSL role and quickly departed after finding out the job was not what they had expected. Many MSLs that did not enjoy the job cited the following as issues for leaving the role: corporate culture, isolation, lack of structure, position was too "salesy", no or little intellectual challenge, and too much travel. Hopefully the information included in this book will provide you with what is needed in order to make sound decisions and truly assess whether or not the MSL role is right for you, if you are considering this opportunity.

First, you may want to consider taking a values assessment. Once you understand what skills you have, and what interests are unique to you, you can then begin to understand whether or not becoming a MSL is right for you. What do *you* value? Knowdell Career Values Card Sort (www.careertrainer.com) is a low cost, excellent product available for values assessment. At the time of printing, there is also a free values card sort assessment available online at: http://www.motivationalinterview.org/library/valuescardsort.pdf. Essentially, this product is a deck of value cards that you can sort into categories of most important to least important to you. Also, you can keep this deck of cards and regularly check your values, as they may change over time. For example, the 'Always Valued' cards chosen by one MSL utilizing a variation of this deck included the following:

- Independence
- Works on the frontier of knowledge
- Change and variety
- Creativity
- Time freedom

Once you have identified the values important to you, you can then explore the skills and abilities necessary for the MSL role. Some are listed below. However, it is important to look at each and every specific job description, role, territory, and company you are considering. Each of these elements can vary widely across the pharmaceutical, biotechnology and medical device industries. The educational qualifications are many times either a PharmD, PhD, or MD. There are rare exceptions to this. Someone with previous experience may be considered for the role without the doctorate.

Skills and abilities specific to the MSL:

- Ability to teach
- Ability to learn at a rapid rate
- Ability to speak to highly technical scientific audiences
- Strong communication skills
- Ability to travel (>50% in many cases)
- Ability to collaborate, persuade and motivate
- Strong computer skills (PowerPoint, Word, Excel, the Internet)
- Ability to seek, understand and match needs of academic thought leaders to company
- Therapeutic area experience
- Knowledge of the clinical trial and drug development process

Examples of Abilities and Skills Necessary for the MSL Role:

- Analyzing
- Assembling
- Building
- Calculating
- Consulting
- Coordinating
- Counseling
- Creativity
- Decision making
- Designing
- Detailing
- Developing
- Diagnosing
- Driving
- Endurance
- Evaluating
- Examining
- Follow through
- Human relations
- Ideas
- Imagining
- Initiating
- Inventing
- Listening
- Managing
- Negotiating
- Observing
- Organizing
- Persuading
- Planning
- Public speaking
- Rendering services
- Researching
- Risk Taking
- Selling
- Showmanship
- Teaching
- Troubleshooting
- Writing

Previously, the 5 core elements of a MSL role were provided. In an ideal case all 5 elements should co-exist. However, these elements vary from team to team, and company to company. The elements can also be affected by the lifecycle of the product(s) that the MSL supports. It is important to ask yourself which of the 5 elements are fun, interesting, and challenging to you, and try to match the position to your skill set and your preferences. For example, some MSL programs have a very small research component. If supporting investigator-initiated trial ideas and concepts is interesting and important to you, then you may wish to consider another company or product. If teaching is not your strong suit, a MSL role with a strong element of managed care formulary presentations would not be blissful.

Part 2: Résumé/CV

Needless to say, the résumé must be easy to read, succinct, and free of errors. However, résumés must do much more, as they convey on paper the skills that a candidate can offer to an organization. Although a résumé/CV is very important, it is argued that the cover letter is a far better tool to utilize for getting your foot in the door for an interview. Also, there are several different styles and layouts of a professional résumé/CV. A single method is presented below; however, there is no best format for a résumé/CV.

After you have taken your values analysis, there is another immensely helpful exercise you can work through in order to create your résumé: a strengths analysis. An easy strengths assessment is the analysis from a book entitled, *Now, Discover Your Strengths*, by Buckingham and Clifton[9]. Another option to study your strengths is to think of 3 separate situations, work or social, where you enjoyed the work you were performing. Sit down and with each scenario, write out a list of all the actions and skills you needed in order to complete the 3 scenarios. If you perform an analysis of three separate situations, you should begin to see a pattern emerge regarding the actions you took around your rewarding work scenarios.

After knowing what abilities you match to the skills of the MSL, you can hi-light these in your résumé. Human Resources (HR) tends to be the first department to screen most résumés. However, HR does not always know the MSL job well. Hence, you want to make it as easy as possible for the HR person to screen your résumé and see your experience quickly, easily, and succinctly. Therefore, we strongly recommend a profile at the top of your résumé. A profile is essentially a top line of your skill set. Example:

PROFILE: Pharmacist with knowledge and experience in medical and clinical affairs, marketing, business development, clinical research and information, adverse experiences, quality assurance, customer problem solving, retail pharmaceutical sales and promotion seeks challenging, rewarding position to expand current skill set.

- Developed a multi-regional territory as a medical science liaison reporting to the medical affairs department in various therapeutic areas for three global pharmaceutical companies.

- Former manager in the clinical operations department of a major biotechnology company with coordination responsibilities for all clinical trials within the pharmaceutical division.

- Practiced and managed pharmacy in retail setting for several different stores in national grocery store chain.

The profile heading should go at the top of the first page immediately after the candidate's name and contact information. Those with little academic experience should have a profile. Those with more of a CV formatted résumé, extensive publications, and/or an extensive work history should consider a profile as well. HR screeners do not have time to read a 50-page synopsis of your work. Also, if you are going for a particular therapeutic area, product, or company, you may want to include your therapeutic area(s) of expertise within the profile.

After the profile, the professional experience should occur in reverse chronological order. Recruiters will recommend both months and years in term for the positions held, but you may wish to consider limiting to years on the résumé. Keep everything bulleted and give your top 5 highlights/responsibilities of the job utilizing the action activity words you utilized heavily in the previous exercise. An example follows:

PROFESSIONAL
EXPERIENCE: PHARMACEUTICAL COMPANY A, Everett, PA

Worldwide Product Safety Coordinator (2000-2001)

- Managed and tracked spontaneous and follow-up adverse events utilizing the Worldwide Adverse Event System database and assisted with quarterly periodic submissions on team-specific medications to the FDA. Disease state responsibilities included: HIV/AIDS, agricultural/veterinary products.

- Updated clinical site listings for adverse experience reporting and co-developed reporting with the Center for Disease Control's *Post Exposure Prophylaxis Program* in conjunction with Pharmaceutical Company B.

After the work history, the following headings or sections should occur (if applicable): education/degrees, internships, professional experience, awards and honors, publications (books, articles, reports, journals), speaking engagements, and professional affiliations. If you have a lot of these, add the category. If not, create them if you have experience in these areas.

If you lack some experience on the research and teaching (public speaking) side and wish to pursue a MSL job, you must increase your expertise in these areas. Can you teach at a community college? Other health care professionals? Peers? Patients? Can you research in areas of outcomes? Can you perform survey research? If these are not of interest to you, you may wish to consider another industry role other than MSL. Finally, do not include your references or the statement 'references available upon request' on your CV or résumé. This is assumed and understood.

An example résumé utilizing the format discussed is provided:

Joe Q. Public, BS, MS
101 Anywhere Street
Louisville, KY 12345
Work: 317.555.4312
Home: 317.555.4313
Fax: 317.555.4315
joeqpub@gmail.com

PROFILE Pharmacologist with 15 years of pharmaceutical industry experience:
- 7 years Medical Science Liaison (OIG/ACCME)
- 3 years clinical trial management (GCPs)
- 3 years non-clinical (GLPs)
- 2 years administrative

- Desires work in biotechnology company and enhance customer-relationship, business development, and/or investor relations skills
- Current Position: Senior Medical Science Liaison with industry training in many therapeutic areas and disease states: Cardiovascular (Atherosclerosis, Peripheral Arterial Disease, Hypertension), Gastroenterology (IBD, IBS, Cystic Fibrosis), Neuroscience (Schizophrenia, Bipolar Disorder), Nephrology (Diabetes), Neurology (Stroke), Pulmonology (Asthma).

PROFESSIONAL
EXPERIENCE XYZ PHARMACEUTICALS, Inc., Indianapolis, IN 46220
Medical Science Liaison (1998-2002); Senior Medical Science Liaison (2002 to present)
- Develop and maintain relationships with 50 Key Opinion Leaders

- Manage grant requests from physicians conducting symposia
- Negotiate cost and launch national continuing medical education programs
- Obtain input from thought-leaders for phase IV post-launch trials
- Create disease state and product-related slide sets: Crohn's Disease, Irritable Bowel Syndrome, Heart Failure, Diabetes, Nephrology, Stroke, Schizophrenia
- Present clinical information to Pharmacy and Therapeutics Committees
- Perform product/competitor analysis; Update team on regulatory status of phase III and IV compounds; Provides strategic direction based on findings
- Maintained clinical knowledge by attending DDW, AHA, APA, university preceptorships
- Excellent knowledge of OIG, ACCME, AMA guidelines
- Possess superb written and verbal communication skills and give frequent oral and written presentations to various groups (internal and external) on the medical and clinical aspects of a given product or program
- Excellent knowledge of Microsoft® suite of applications
- Geography: Indiana, Ohio, and Kentucky.

BIOCOR, Inc., Malvern, PA 19355
Clinical Research Associate II (3/1997 to 7/1998)
- Oversaw and guided clinical research organization during Phase III trial according to SOPs/GCPs of drug A (trade name®) and drug B (trade name®)

- Conducted pre-study site visits; evaluated potential clinical investigative sites
- Conducted initiation, interim, and close-out visits to assure that the clinical investigation was being executed per protocol, GCP, and federal regulatory guidelines
- Conducted audits of enrollment sites on company's behalf prior to FDA audits
- Wrote amendments, protocol instructions, newsletters, monitoring guidelines and status reports

ABC PHARMA, Inc., Wilmington, DE
Clinical Research Associate I (6/1996 to 3/1997)
- Monitored 3 clinical pulmonary/asthma protocols across North America according to SOPs/GCPs.
- Investigated and updated visit status of seventy-two sites in a four-year-old clinical trial.
- Upon recruitment, followed supervisor to Company C.

THE BIOTECHNOLOGY COMPANY, Inc., Princeton, NJ 08540
Pre-Clinical Research Associate (10/1992 to 6/1996)
- Conducted and monitored acute, subchronic, and biodistribution experiments utilizing liposomal drug delivery systems for antifungal and antineoplastic compounds, i.e., liposomal doxorubicin, adriamycin, paclitaxel, amphotericin B, cyclosporine

- Trained biopharmaceutical support staff to perform radioactive/toxicologic studies, interpret toxicologic trends, dose animals, and comply with federal and state regulations for pharmacokinetic studies.

PROFESSIONAL INTERNSHIPS

Drug Company D., West Point, PA (1992)
Developmental and Reproductive Toxicology Intern

PHILADELPHIA COLLEGE OF PHARMACY AND SCIENCE (1990 to 1992)
Oncology/Toxicology Intern

PROFESSIONAL AWARDS

Team Spirit Award (2005)
XYZ Company Star Award (2003)
Innovation Award (Honorable Mention, 2002)
Celebration Award (2000-2004)

EDUCATION

BS in Psychology, Villanova University, Villanova, PA (1986)
BS in Pharmacology and Toxicology, Philadelphia College of Pharmacy, Philadelphia, PA (1992)
MS in Quality Assurance/Regulatory Affairs, Temple University, Philadelphia, PA (2002)
Certificate in Competitive Intelligence, Drexel University, Philadelphia, PA (2004)

PROFESSIONAL POSTERS & PRESENTATIONS

"Competitive Intelligence 101: A Starter's Guide." Workshop at the Drug Information Association Annual Meeting, 2005, Washington, DC.

"The Epidemiology of Stroke." CME Presentation, ACME Hospital, January, 2004.

RÉSUMÉ EXAMPLE # 2:

Anastasia Smith, MBA, PharmD
2345 Any Street
Indianapolis, IN 46220
(317) 222-3333
(317) 222-3334 fax
asmith@gmail.com

PROFILE: **Pharmacist** with knowledge and experience in medical and clinical affairs, marketing, business development, clinical research and information, adverse experiences, quality assurance, customer problem solving, retail pharmaceutical sales and promotion seeks challenging, rewarding position to expand current skill set.

- Developed a multi-regional territory as a medical science liaison reporting to the medical affairs department in various therapeutic areas for three global pharmaceutical companies.

- Former manager in the clinical operations department of a major biotechnology company with coordination responsibilities for all clinical trials within the pharmaceutical division.

- Practiced & managed pharmacy in retail setting for several different stores in national grocery store chain.

THERAPEUTIC AREA EXPERIENCE:

Central Nervous System: Sleep medicine, schizophrenia
Women's Health: Osteoporosis, hormone replacement therapy
Gastroenterology: Ulcerative Colitis, Crohn's disease (Anti-TNF)
Cardiovascular: Percutaneous Coronary Intervention (GP IIb/IIIa)
AIDS/HIV: (Protease Inhibitor)

PROFESSIONAL EXPERIENCE:

DRUG COMPANY X, INC. (Philadelphia, PA)

Medical Science Liaison, CNS (2004-present)

- Establish, develop, and maintain relationships with current opinion and thought leaders within a multi-state territory (Michigan, Ohio, Indiana and Kentucky). Promote awareness of medical developments and issues related to insomnia/sleep medicine to influential members of the medical community.

- Facilitate various educational and health management programs. Serve as a liaison in identifying and fostering investigator-initiated trials and CME opportunities. Support scientific exchange efforts by presenting clinical research data via unsolicited request to healthcare providers in managed care and clinical practice environments.

DRUG COMPANY Y, Cincinnati, OH

Senior Medical Science Liaison, Osteoporosis and Gastroenterology (2001-2004)

- Developed relationships with opinion and thought leaders within a multi-state territory (Indiana/Illinois). Facilitated opinion leader speaker training, regional consultant meetings, and MSL training projects during tenure.
- Promoted from MSL to senior MSL during assignment.

PHARMA AND COMPANY, US Affiliate, Indianapolis, IN

DrugQ Alignment Project Associate, DrugQ Brand Team (2001)

- Lead the DrugQ Alignment Project, an organizational design and change management process focused on the implementation of a new brand position and strategy across targeted customer segments.

Marketing Plans Associate, DrugQ Brand Team (2000-2001)

- Managed content on multiple advisory boards for DrugQ, including bipolar and schizophrenia. Coordinated data between medical, legal, vendor, and lecture bureau to revamp medical slide kits for advisors, consultants, and lecture bureau members.

- Managed speakers regarding closed symposia. Partnered with medical/professional organizations. Built relationships with external advisors, consultants, and speakers. Bridged between medical and the brand team, gathering new medical/clinical content for medical meetings.

THE RETAIL QUICKMART CHAIN,
Central Marketing Area, Indianapolis, IN

Pharmacy Intern, Staff Pharmacist, Pharmacy Manager (1992-1996; 1998; 1999-2000)

- Promoted from pharmacy intern to staff pharmacist, and to pharmacy manager with national pharmacy/grocery chain. As a staff pharmacist and a pharmacy manager, hired and trained all pharmacy support staff on computer software, customer service techniques, and day-to-day operations. Instructed and served as preceptor for fifth year BS and sixth year PharmD externship students. Taught students how to provide drug information and counseling, integrate drug utilization review, and comply with federal and state pharmaceutical laws.

- Recruited a local psychiatric hospital and negotiated an exclusive contract to supply drug information, consulting, and pharmaceuticals. Recommended hospital formulary additions and substitutions.

SURE WAY PHARMACEUTICALS, Marietta, GA

Professional Services Associate (MSL), Women's Health (1998-99)

- Established, developed, and maintained relationships with opinion and thought leaders. Promoted awareness of medical developments and issues related to company products to influential members of the medical community.

- Facilitated various educational and health management programs. Served as a liaison in cultivating and obtaining speakers for various peer-influence, grand rounds, and certified medical education forums. Supported managed care efforts by presenting clinical research information to pharmacists at targeted PBMs, HMOs, and PPOs.

- Educated professional field sales representatives by providing clinical presentations at company meetings. Managed clinical information and technical support for the professional field representatives and targeted physicians within the regions.

CENTEON, INC., Malvern, PA

Clinical Research Assistant, Clinical Operations Associate (1997-98)

- Forecasted and maintained clinical drug and supply inventories worldwide, including ventures with marketing partners, independent clinical sites, and overseas company facilities. Headed an internal multi-department task force to optimize clinical material forecasting, handling, and billing for both internal and partnered clinical trials for pipeline products.

- Coordinated site, legal, and financial contracts on cardiovascular and gastroenterolgy, and rheumatology protocols. Acted as liaison between sites and corporate attorneys for all clinical contracting issues. Managed payment allocations to clinical sites. Created and updated monthly trial enrollment status reports.

PHARMA & COMPANY, Blue Bell, PA

Worldwide Product Safety & Epidemiology Coordinator (1996-97)

- Managed and tracked spontaneous and follow-up adverse events utilizing the adverse event database and assisted with quarterly periodic submissions on team-specific medications to the FDA. Disease state responsibilities included: HIV/AIDS, agricultural/veterinary products.

- Updated clinical site listings for adverse experience reporting and assisted with the Center for Disease Control's *Post Exposure Prophylaxis Program* in conjunction with Pharmaceutical Company R.

EDUCATION:	PharmD, Purdue University (2005)
	MBA, Marketing, Indiana University (2001)
	BS, Pharmacy, Purdue University, Indianapolis, IN (1994)
MEMBERSHIPS, AWARDS & COURSEWORK	Licensed to practice pharmacy in Indiana (1994-present)
	Introduction to Dreamweaver – IUPUI and macromedia.com (2006)
	Drug Information Association – Member (2005)
	Strategic Brand Management, Tuck School of Business, Dartmouth College (2001)
	GCP Training Course, Pharmaceutical Education & Research Institute, Inc., Philadelphia, PA (1997)
	CRA I Seminar, Drug Information Association, Philadelphia, PA (1997)
PROFESSIONAL PRESENTATIONS POSTERS & PUBLICATIONS:	"MSL Responsibilities at Conferences." Panel discussion at Exl Pharma's MSL Research & Best Practices Seminar, December 7-8, 2005, LaJolla, CA.
	"Best Practices in Retaining and Training MSL Teams." Workshop Exl Pharma's MSL Research & Education Best Practices Seminar, December 7-8, 2005, LaJolla, CA.
	"Field-based MSL Job Satisfaction: Three year results." Poster 222, presented at the American College of Clinical Pharmacy 2005 Annual Meeting, Oct 23-26, San Francisco, CA.

* *NOTE: all information in resume examples was fictitious.*

Part 3: The Cover Letter

The Cover Letter is the most underutilized and helpful document known to mankind. If you structure your cover letter correctly, it will more likely get you an interview over the CV or résumé more times than not. Why? A properly structured cover letter will give the HR professional a checklist of the jobs' qualifications versus your individual skills in one succinct page.

An executive briefing letter[10] is highly recommended, particularly in two situations. First, it may be optimal for a healthcare professional seeking a MSL position without previous MSL experience. Second, the executive briefing cover letter may more effectively sell the candidate to the human resources department when not working with a recruiter to do the sales and marketing on their behalf. It requires a little more customization and a job description from the company with which you are seeking a position, but it makes everyone's lives easier on the hiring side. Here is what you need in order to develop a customized match needs cover letter:

1. Obtain a copy of the job description for the job you seek. The following will be used for the sample (note that the company and drug names are fictitious):

XYZ Pharma*, based in the Miami, FL area, has spent more than a decade building a world-class research and development organization. **Wonderdrug**™ is currently available in approximately 100 countries, including Australia, Brazil, France, Mexico, Canada, the United States and countries throughout Europe. XYZ Pharma is committed to advancing science in the areas of protein based and small molecule therapeutics.

If you are a Medical Science Liaison with significant experience and want to work in a fast-paced environment of a growing company, then consider XYZ Pharma.

Responsibilities:

The MSL is responsible for furthering the medical community's scientific knowledge of XYZ PHARMA products by providing up-to-date medical information from XYZ PHARMA to local and national thought leaders. The MSL will also provide an avenue for thought leaders to propose additional research ideas for established uses as well as additional indications. The job responsibilities include:

- Develop relationships with thought leaders in specified geographical regions to allow rapid dissemination of scientific data on XYZ PHARMA products.
- Facilitate investigator-initiated research by discussing research ideas with potential investigators and ensuring appropriate transmittal of research proposals to appropriate entities.
- Maintain up-to-date scientific knowledge of the XYZ PHARMA compounds and therapeutic areas by attending appropriate scientific meetings and by literature searches, and library research.
- Respond to specific requests for medical information regarding XYZ PHARMA compounds.

- Support communication between external physicians and the internal physicians and provide competitive information to the product teams.

Job Requirements:

- Excellent interpersonal and communication skills (both oral and written).
- Significant experience in creating and delivering scientific or medical presentations.
- MD, PharmD, or PhD in biology or other related science
- Excellent self-management skills.
- Strong interest and motivation to provide comprehensive support for community medical education.
- Demonstrated ability to work collaboratively in a team environment.
- Understanding and focus on the needs of health care professionals.
- Prior medical affairs or clinical research experience preferred.
- Ability to use field-based electronic or other communication tools is essential.

This position will be based in Indiana. The position will require significant travel by automobile and air.

XYZ Pharma offers a highly competitive compensation package and a friendly, collaborative culture that values personal initiative and professional achievement. EOE

Fictitious.

2. Go through the job description above and hi-light (or in our case, underline) all the skills/requirements necessary for the job. As you can see, this has already been completed in this example.

3. Next, take each skill hi-lighted in the job description and put them on the left hand column of your cover letter under the heading "Your Requirements" and on the Right, "My qualifications." Then match their needs with your unique skill set. Example as follows:

XYZ Pharma
PO Box 1234
Miami, FL zip code

RE: Medical Science Liaison, Indiana

In response to your advertisement on XYZPharma.com as of (date), I have listed some of my qualifications to parallel your stated requirements:

YOUR REQUIREMENTS	MY QUALIFICATIONS
MD, PharmD, or PhD with significant medical experience.	MD with board certification in (disease state) and 10 years' experience.
Ability to Develop Relationships with Thought Leaders	Currently colleagues with several thought leaders within the region. Trained at major institution within territory, Indiana University.
Facilitation of Investigator-initiated research	Worked with medical teams to adapt SOPs and procedures for on site medical injuries. Researched XYZ

	during tenure at company Y and published it at Z.
Maintain up-to-date knowledge in therapeutic area by attending medical conferences and literature reviews.	Attend regularly (therapeutic) meetings. Member/speaker at Indiana medical society. Performed literature review for last 4 wellness programs at facility.

...and so on...keep going down the skills/requirements in the order that they appear, and match to your skills...

Contacts/Networking Strategies

Once you have your values/strengths assessed, your CV polished and your cover letter in order, it is time to get out your personal Rolodex™ and begin with people you know. Who do you know within the industry? Start contacting them.

Secondly, it is important to think about which companies you wish to target for a medical liaison position. What therapeutic area(s) are you qualified to work within? Another issue to consider is pay and geography. As someone entering the pharmaceutical industry with little industry experience, ask yourself if you are willing to take a pay cut in order to become a MSL?

Typical salaries can range from 80-140 thousand US dollars based upon experience. Many physicians struggle with pay issues as they can often make more salary working within private practice. A PharmD may make similar salary to retail pharmacy or more as a MSL than in academic pharmacy or hospital pharmacy. However, the MSL role has benefits such as company cars, stock options, stock purchase plans, and other corporate perks that many healthcare providers do not receive in their practice settings.

Are you willing to relocate? If the MSL job you want is within the therapeutic area of expertise you desire but the position is on the opposite side of the country, are you willing to consider it? How will it impact your family?

The industry-naïve MSL may struggle with developing contacts. Recruiters often place many MSLs within roles before they are even posted on the Internet. However, keep in mind that if you have never been an industry MSL, you probably will not get far with recruiters. They are paid between 25-30% of a MSL recruit's base salary (yes, $25-45K) in order to place a MSL and therefore prefer to bring highly experienced and qualified candidates to the company.

You might first try looking at MSL contract companies that sometimes take industry-naïve candidates for MSL positions. Some include The Medical Affairs Company (www.themedicalaffairscompany.com), Innovex (now part of Quintlies - http://www.innovex.com/ServicesSolutions/IMC.htm), or Science Oriented Solutions (SOS, www.medicalaffairs.com). Some companies have MSL positions and also occasionally offer clinical liaison positions, which is more of a clinical educator and a bridge between sales and the MSL. Many Clinical Educators or Clinical Liaisons are advanced nurses or RPh/BS level pharmacists. These companies contract with bigger pharmaceutical companies to find talent for them in lieu of a recruiter. If you were hired, you would technically work for either Innovex or SOS until your contract runs out or you are given an offer by the company that SOS or Innovex contracted with to become a full time employee.

Do these firms pay less? Sometimes yes, sometimes no. However, they are more willing to consider candidates without a lot of industry experience, provided they have a doctorate and/or the clinical experience necessary to speak the MSL language. (For more on contracting MSL positions, see the interview in the book with Kyle Kennedy of The Medical Affairs Company in chapter 9.)

The second best route, after exhausting your personal contacts and considering the routes listed above is to apply online through the job posting websites, such as monster.com, medzilla.com, hotjobs.com, etc. These sites often post many MSL jobs. Keep in mind that you should try and utilize the executive cover letter whenever possible in this situation, as it may allow you to get past the HR screen and get you to the hiring manager for a review.

Getting Into the MSL Role: A Recruiter's Perspective - Interview with Tony Beachler

Tony Beachler always wanted to get into the pharmaceutical industry. In fact, the reason he became a recruiter was because he would be dealing with the pharmaceutical industry and thought he would just find a job in the fall of 1996. That was the golden age of pharmaceuticals. He was enjoying the recruiting aspect and did that from 1996 to 2000 with a Grand Rapids, Michigan firm. He did not get involved in the MSL recruiting role until 1999. He stumbled into the MSL recruiting function through another recruiter that joined his firm. Once he started with MSLs, he began to understand their needs were different from pharmaceutical sales. He later started his own recruiting company called BioPharma WorldWide and currently runs BPWW out of his headquarters in Grand Rapids, Michigan. Tony can be reached at: his website, www.bpww.com, via email at: tony@bpww.com, or via phone at: (616) 459-6000.

What are the differences between recruiting a sales representative and a MSL?

A lot derives from physicians and key opinion leaders. The clinical nature and the education of the MSL roles are big differentiators. There are big differences between a MSL 10 years ago versus 5 years ago and even today. Now more than ever, a MSL must have a personality and people skills beyond just pure science. The companies are a lot more stringent on doctorates, and they still must have people skills on top of the baseline doctorate. I still have some old school MSL managers that are non-doctorates, but they are not able to easily hire other non-doctorate MSLs, and most likely cannot obtain work as a MSL through a recruiter. At least 95% of all MSL opportunities through a recruiter require a baseline doctorate.

For people that don't have a doctorate and want a MSL job, how can they go about seeking an MSL opportunity beyond working with a recruiter?

It's all about networking and not just a résumé. It's all about who you know, and you have to be more creative. A lot of managers today want the out of the box MSL. If I were the MSL-to-be and I found out who a MSL manager is through my network and moved my résumé onto the hiring manager's desk via 3 or 4 different routes, the managers tend to be impressed with this.

It sounds as though it's similar to getting in to see that difficult to see thought leader.

Absolutely. Yes. They don't care how the candidates get to the managers, as long as they get there.

You mentioned it's more than just a résumé that gets a MSL an interview. Can you elaborate?

A great place to start is with sales people. Sales people love to talk. If you ask them who their MSL is, some may look puzzled as they never worked with one before, but most sales people know their MSL. Get a name, and if you don't have an office or cell number, 9 out of 10 companies have voicemail. Get the 800 number for corporate voicemail and dial by directory. Or call corporate and ask who is the MSL for a certain state. Call and introduce yourself. Most people like to help people, and if nothing else, the MSL may be able to let you know more about what the job means at their particular company. The 'about us' section of every company's website is a great place to start too. The higher up you leave a message, the greater the possibility it will trickle down to the appropriate hiring manager.

What about corporate human resources (HR)?

In today's very aggressive challenging workplace, a lot of companies have vendor lists for recruiters. Recruiters don't work for every company. The companies pay the recruiters for placement of candidates, unlike a lot of industries. Recruiters work for selected companies. Sometimes if I have a stellar candidate, I would contact a specific company and market that candidate to a specific company.

What does a great CV look like for MSLs?

Not too many job jumps. A lot of publications. A doctorate degree. Current or recent MSL history, and not as an intern. Too many job leaps are a red flag. Therapeutic classes don't really matter. I will contact a MSL and start a dialogue to see what they want to be doing. Today's MSLs have an open mind as to what's out there in therapeutics. Getting too specialized in one area for example, may be narrowing one's opportunities. As long as a MSL has good baseline knowledge, a great can-do attitude, is personable, has good eye contact and smile - one will go a long way through the hiring process if not be hired. Attitude is really important. Smaller companies are also narrowing territory assignments as well. It is just a matter of aligning yourself with a recruiter to keep you informed of what is going on in the industry. As a recruiter, I depend on the MSLs very heavily as well to hear what is going on in the industry. We can then keep a good temperature of what's going on within the industry. Finally, timing of everything is key for new opportunities.

How do you want MSLs to work with you as the recruiter?

First and foremost, I need MSLs to communicate with me. I'm not naive, I realize MSLs will work with several recruiters. We are here to work for the companies, but good recruiters will work heavily for candidates that are open and communicative.

How can a MSL sniff out a bad recruiter?

There are a lot of non-professional recruiters out there. Use your gut - you can tell whether or not the recruiter is in it for the dollars, or if they are in it to help the candidates. I always try to tell people out of the gate that good people should stay 3-5 years and then make a change, then repeat. The industry average for a MSL now for length of tenure is 18-24 months. You talk to some hiring managers that think a MSL in a position for 5-7 years may be a bad thing. I think that's where today's MSL must have sales skills. A graduate PharmD today should consider pharmaceutical sales as a first job out of college.

Why?

MSLs, just like everyone else, are salespeople. They are selling science. If you give MSLs the tools with training on how to open an appointment, listening skills, becoming a partner, solve problems, and close, the sky is the limit for anyone. Where do you want to go? A doctorate level scientist that understands sales can go several places - regulatory affairs, clinical operations, marketing, etc. The career path will be broadened and the dollar amounts will be too.

Sales is a dirty word to a lot of MSLs and scientists...it turns them off.

Yes, but I see a paradigm shift between MSLs hired 5 years ago versus today.

What about the young graduate going in house to medical information as a first job out of school? Any other in house routes?

That is difficult. If one moves in house - the good news is you can network internally and meet the right people. The bad news is that you may make yourself too valuable to move out into the field, then you can be stuck.

When a candidate interviews - how can they figure out if they may have a good or bad manager?

First impressions are very accurate. I want all MSLs to know that they can interview and they don't have to take the job. You as the MSL have to be as sold on the company and the manager as they are about you the candidate. Ask the tough questions. Ask the hiring manager about their style. Ask why the position is open. Ask them what their competitors would say about them. Ask them where they compare to their competitors. If there is an older product or a product in a marketplace that is near the bottom, the MSL needs to consider that level. Do you want to work for this organization? MSLs are of great value to a company. There are a lot more positions out there than qualified candidates. The candidates that are qualified can take their time and study what is available.

What are the motivators for MSLs?

It's not always about money. But the MSL needs to be careful of the culture of the company they are working for potentially. To understand the culture, I would ask myself as a MSL - do I see myself here long term? Is this a stepping-stone to get me to another company? As long as you are honest with yourself, there is nothing wrong with knowing that. I would also talk to the sales people in the company you may work for, and also talk to other recruiters about the company. Cafepharma.com has good and bad points, but people are candid on that site.

What about cover letters?

Not really important if you are working with a recruiter. The recruiter's job is to literally be the cover letter for the candidate to the company. Our job is to secure the interview, then the rest is up to you and the chemistry. I help candidates through the entire recruiting process.

How do you help through the entire process?

We help provide the story behind the story. Why the position is open, what the company's culture is like, etc. Before I even pick up the phone to find candidates, I know whether or not I would go work for a company....and if I wouldn't, I would be careful about how I sell the position. As a recruiter, I need to go in and place a good recruit call to a happy MSL in order for them to call me back. If the MSL has been 3 years or more in one position, they owe it to themselves to explore other opportunities. Nine out of 10 times there really isn't much of a difference in total compensation for MSLs. There is a 10 percent increase in pay, typically.

Some may even take a pay cut for a smaller territory.

True. If someone is covering San Francisco all the way to Texas and the Southwest, they may take a pay cut for just one state.

If a MSL works with a couple of recruiters and switches jobs, should they update their CV and send it out to recruiters?

That would be great, but that rarely happens. Once a person accepts a position it is a card here and there or a phone call, but they feel guilty if they contact previous recruiters if someone new placed them in a new job. They shouldn't feel badly about it, but if they can keep in touch, opportunities may be better with me again in 3-5 years when they are seeking other opportunities.

What types of missteps are candidates making?

MSLs need to remember the little things. Handwritten notes and phone calls are important. Candidates are not doing this as much as they could. I think is a place where really exceptional MSLs separate from what I call 'skaters.' They don't work really hard, tread water, stay off the radar screen and do their jobs, but they have lost their passion. The little things like sending a hand written thank you note or knowing about their customers as people is key to bonding and forging good relationships. A lot of the products are similar. It comes down to why do I want to work with this MSL versus another. Will that person be reliable?

What can MSLs do to move on from the MSL job?

I've seen MSLs that were with big pharma leave to start their own companies or start up MSL teams. It is really important for MSLs to network with each other as well as their thought leaders and former colleagues. Networking doesn't mean you are unhappy. It means you are connected to others and you know what is going on within your industry outside of your company. If something new comes along, great. But opportunities will never come to you if you don't work with recruiters. We know we may catch MSLs at bad times, but a large percentage of MSLs do return my calls. It takes 30 seconds to call back. If the MSL can share with me what their utopia looks like, that would be great, but good recruiters already ask this question. The MSL community is very close knit and difficult to penetrate. Every good individual is a free agent. Every great person wants to get to a winning company or team that would give them flexibility freedom or great thought leaders, or more money, whatever....I don't know what that is unless they tell me. But if they do, it's a win win-win situation. The opportunities for MSLs are increasing each year, which is great news for MSLs.

Ten Questions to Consider if the MSL Role is Right For You:

1. Do you like working on the forefront of scientific knowledge?
2. Do you enjoy reading and interpreting research?
3. Do you like working in a creative, gray arena?
4. Do you enjoy ideas, discussion, and scientific debate?
5. Are you persuasive?
6. Do you enjoy giving presentations?
7. Do you like to travel? If so, how (by plane, car, etc.)
8. Can you work independently?
9. Can you work from home?
10. Do you enjoy learning?

Part II:
Perfecting the Art of Liaising

Chapter 4:
Starting off as a MSL

Congratulations, you got the MSL job and now you are off and running! But, where are you going? Here are some tips and ideas for your first year as a MSL, which can be a challenge to many. The first year can make or break many professionals that choose the MSL role. It can be isolating, vague, and frustrating. It can also be professionally rewarding, provide autonomy, and the ability to be creative. As previously stated, the role is definitely not for everyone.

Here are ten tips on your first year as a MSL:

1. Get organized: One of the greatest perks and curses of the MSL job is that you are based at home. That means the work is always there. It is imperative that you get your space organized. Your company is going to send you about fifty boxes of stuff. You not only need good office space, but you'll also need good storage space. You should really try and establish your office workspace in one room. Try and ensure it has a door on it too: the more you can seal off the space after a long hard week of working and travel, the more sane you'll be and less tempted you'll be to keep working on the weekends.

Some companies will supply the new MSL employee with an allowance to buy office hardware, items such as a desk, a filing cabinet, bookshelves, and an office chair. Other companies will not. If you are a first time MSL and don't have this equipment at home, it may be worth asking if this allowance is offered (once you have the offer of employment in your hand, not before). If the company does not offer an allowance, be prepared to spend some of your own money to set up office space. If your company does not offer this, you could always ask for a sign on bonus to set up your office (again, not until you have offer in hand).

The items headquarters is going to send the MSL can include: books (usually the large medical textbooks, which require a heavy duty bookshelf), a computer (usually a laptop and a docking station/monitor). If you are limited on space, just take the laptop and return the docking station. A printer/fax/scanner or several pieces of equipment come to the office. These are usually large and a necessity. You'll also receive binders, a lot of binders, ranging on topics from corporate and regulatory policies to training on a disease state and on the drugs the MSL will be working with. You may also receive corporate stationary.

Finally, the MSL may also receive a hand held device, either a cell phone, an all-in-one device (such as a Blackberry®) or in some cases, both. Some companies will pay for your cell phone and the monthly charges, some will only pay for the monthly charges, and some will limit the amount the MSL can expense per month. It is important to ask what phone charges are not covered by the company. Some MSLs also have a traditional office landline, a fax line, and a cell phone line. If you are a new MSL, try asking your manager for two things at your start date: 1. A list of all the items that will be on the doorstep eventually and 2. The name of a buddy or a mentor MSL on your team that can advise you about how to set up your office. It is really important to keep an inventory of the items that were sent to you from the company, because one day you may need to return all of it.

2. Get a buddy: The new MSL needs someone other than the boss to ask questions of, get advice from, and mentor them throughout the first year. Ask your manager for mentor suggestions. If the manager does not respond, the MSL can find their own personal buddy or mentor through less formal means. This is one of the most critical steps for a first time MSL. The job can be isolating, particularly in the first three months when the new MSL is not yet seeing customers and is knee deep in learning, reading and studying the therapeutics of the new assignment.

The interviewing MSL should also ask about what mechanisms (informal and formal) exist for training and development at the company they are interviewing. Larger companies tend to have more formal training programs, which are critical to the first time MSL. Other companies, smaller in size or in start up mode, tend to have less formal training programs. If the individual MSL is a self-motivated learner, this may not be an issue. However, if the new MSL feels overwhelmed and has no one to discuss the mechanics of the job with, the MSL will end up feeling isolated and/or potentially disliking the job. If possible, also find someone that started at the same time. A buddy can share their study tips and office set up with you; a mentor (or a MSL that's been there for 5 years) won't be at the same level. If you can find someone to commiserate with during the training process, learning in a group or team is a lot easier than on one's own.

3. Be like Switzerland: Companies, especially larger companies, can be politically charged. Office politics are somewhat removed from the daily life of a MSL, but one that works for any organization can never truly escape from the curses of gossip, backstabbing, and toxic work environments. The most important advice for the newer MSL is this: be like Switzerland. Remain neutral. Do not take sides. Certainly you can stand back and listen to all opposing views, but it is truly important to make your own call and judgments after you have all the facts. Taking sides early on can be political suicide to any employee. Do not get caught up in the temptation of the political environment in the office. Besides, you will have your hands full trying to learn the job and your territory, and that is enough to focus on for at least the first year of your assignment. Similarly, there can be rivalries between institutions within your geography. These rules apply here as well: learn the facts, take everything in, and remain neutral.

4. Discover the lay of your land: One geographic assignment will not be equal to another. Each university, hospital, and thought leader will have their own unique set of requests, policies, and people. If the new MSL has a multi-state territory, knowing where to start can be overwhelming. If a company provides a starting point and/or there was a previous MSL assigned to the territory, the job becomes much easier. Even if someone in the company used to cover the assignment in a different therapeutic area, it is still worth picking the individual's brain to understand the dynamics of the territory.

For example, one major academic institution kicks out anyone associated with a pharmaceutical or biotechnology company. Just like famous brands of shoes or soft drinks, academic centers have their own unique set of rules and cultural eccentricities. Study and learn them, for one should know the rules of engagement before stepping into a potential minefield. Pleading ignorance once may work, but beyond that could spell trouble.

5. Don't take things personally: The MSL may have a great list of thought leaders ready to go, they may study and know their therapeutic area(s) like the back of their hand, but doors may never be opened, email may never be answered, appointments never made, or the MSL may experience negativity due to a previous person that did damage to their company's reputation or name. Don't take it personally. Control what you can control: yourself. Under promise, over deliver, and do your personal best. If the person on the other end sees you are acting with integrity and doing the best that you can with what you are given, eventually, you will be respected. The MSL's job is about long-term relationships, not short term. If a new MSL starts taking things personally, the job is going to get old, fast. Realize that dealing with previous company baggage is just part of the job.

6. Join a club, society or professional organization: Having and growing your professional network is critically important no matter what profession you work within, but it's especially critical to the MSL. There is a list in Appendix A of this book for professional societies to consider joining. During your first year as MSL, you may consider joining a society within your therapeutic area, an industry wide organization, and/or an organization, based upon your professional background (i.e. pharmacy). Three organizations tend to be a good number and provide enough variety for the MSL to understand the dynamics going on within their profession, their industry, and their therapeutic area. Joining such organizations exposes you to other possible mentors within industry and possible future career alternatives. Many companies will also support the MSL's membership fees to these organizations. Ask your manager and join the groups you are interested in learning more about.

7. Remember where you came from: Start with the easiest part of the job in terms of networking, who you already know. Try not to reinvent the wheel. Who do you already know in the therapeutic area that is teaching, doing research or lecturing on your topic? Tap into your pre-existing networking and grow your contacts from the people you already know.

8. Widen your scope: Don't just call on physician thought leaders. There are a lot of other healthcare professionals, patient advocacy groups and other types of professionals that the MSL can collaborate with on projects. Each thought leader has a support staff behind him or her as well, clinical research nurses (certified clinical research coordinators, CCRCs), statisticians, co-investigators, fellows, residents, pharmacists, physician assistants, occupational therapists, respiratory therapists, etc. What allied healthcare professionals work within your particular therapeutic area? Look at state hospital associations if you work with products on an in-patient basis. Look for patient advocacy groups for outpatient therapeutic areas. Some therapeutic areas cover both. Also check out the American Society or International Society of your disease state chapter (found in Appendix A). It never hurts to broaden your network.

9. Solicit best practices from your colleagues: Make mentors not only of one of your colleagues, but all of them. Everyone has expertise in at least one area. What are the individuals on your team known for? What are their strengths? Ask your buddy or mentor about other people on the team to get education and ideas from, and make this a habit. Pick up the phone and call your colleagues, a simple conversation about what is going on can lead to great collaborative ideas.

10. Attend programs your company provided unrestricted educational grants toward: Many companies now have a centralized repository for the unrestricted educational grant (UEG) process. However, people at the end of the day connect with individuals. Company representatives can attend continuing medical education programs. It is important for the MSL to attend these events. Not only does it provide a face of the company to the institution holding the event, but also the individual MSL can learn more about the disease state, institution, and speakers merely by attending the events. Just be sure to clarify your company's policy on attendance beforehand.

Miscellaneous for the New MSL

A word on disease state education: it will take awhile to become comfortable holding a conversation in a new disease state. Even if you learned a lot about it in school, what was learned was probably the tip of the iceberg. If the individual came from a bench or PhD background, it may take awhile for the MSL to get comfortable providing clinical relevance regarding the science to clinicians. Regardless, it usually a minimum of 3-4 months to ramp up on clinical data and therapeutic/disease state information before the new MSL even visits with a customer. Even after the first face-to-face interview, the new MSL is not going to know everything regarding a disease state. Not knowing an answer once is fine, as long as the MSL looks up the answer and knows it when asked a second time. Get comfortable saying, "I don't know", you'll probably be doing it often. Just remember to follow up when you find the answer. Some companies have a very formal, highly structured training program for MSLs. Some training programs can last a month. Information overload can occur. It takes time and live appointments within one's territory to get fluent with the data. Many companies will also send you to general training courses, such as facilitator or speaker training. Identify your weak areas and ask to attend those specific courses if they are not mandatory.

Joint calls: A MSL manager will join the MSL in their individual territories usually several times per year. It may be advantageous to have your manager join you in the territory early on to better understand how a good appointment looks and feels. Some companies have new MSLs work with more experienced MSLs in either MSL territory. The more joint calls the new MSL can see and participate in, the better. This also brings up another type of joint call: with the MSL and the pharmaceutical sales representative. In the past, this was common. The experienced representative could introduce the MSL to the thought leader and the MSL could then take the rest of the appointment time. However, due to regulatory guidelines, this practice of joint calls has been left in the past.

Erin's List of Over 20 places to find Thought Leaders

Chances are, your company will already have at least an international/national list of thought leaders, as they may already have some on advisory boards, or involved in early phase research. However, there is an art to finding all the thought leaders in your region. Some thought leaders are very obvious, some are less obvious. Some may be up and coming to the profession and therapeutic area, and some may be new to your region. Below are some suggestions to try if you are either stuck or are just beginning in your territory:

1. The person you replaced, if possible. If you can chat with your previous MSL, it is definitely worth the time.

2. Academic university hospital websites. A departmental grand rounds or journal club is a fabulous place to meet researchers in a particular area. Meet with the department chair to see if you can attend these meetings as an observer.

3. Do a Pubmed search (www.pubmed.gov) in your therapeutic area. Which doctors keep popping up with publications? Do a search on them. Look at their co-authors for other ideas.

4. Asking other thought leaders for referrals. Who would they like to hear speak?

5. Same as #4, but ask about counterparts at their institution in preclinical research and/or animal research.

6. Same as #4, but when you meet with your first round of thought leaders, have a list of tangential therapeutic areas ready to ask them about.

7. Clinicaltrials.gov for researchers in your area.

8. Attending medical meetings and listening to local/regional/national health care speakers, (i.e., state medical society meetings, CME programs, etc.).

9. Your company's investigator lists.

10. Your thought leaders' fellows.

11. Ask your sales representatives.

12. Ask directors of patient advocacy groups about their medical advisors.

13. Attend local hospital and university professional network events.

14. Look for directors or officers of national and state medical organizations.

15. Locate thought leaders/physicians on the American Medical Association's doctor finder website: http://webapps.ama-assn.org/doctorfinder/.

16. Network, network, network. Try Linkedin.com or ecadamey.com. Online and offline networking is essential to the MSL function.

17. Ask study coordinators and administrative people in the department's office and the CME office. Obtain old and new agendas to medical meetings from the CME office if you can.

18. If you are having trouble seeing a particular thought leader, ask if another who knows you to make an introduction.

19. Watch your major metropolitan media channels to see whom they interview for a particular disease state or therapeutic area.

20. If you find a thought leader in another region that one of your thought leaders works with, share that contact with your MSL counterpart. The favor may be returned.

21. Medical textbooks - go through a table of contents and find out if there are authors in your region.

22. Same as #21, but with journals in your particular therapeutic area(s).

Avant-Garde Customers: Ask any MSL about who their customers are, and you will often receive the typical response of "thought leaders". However, the savvy or more experienced MSL understands that while thought leaders may be the MSL's primary customer, there are many others. This is particularly true for products the MSL is supporting either very early in the product lifecycle (perhaps even before product launch). Here is a list of potential customers off the beaten path with which the MSL may consider working:

Bench Scientists: Many companies overlook the opportunity to understand the biochemistry of molecules at the cellular and sub-cellular level, which is a missed opportunity. The FDA approves some drugs, yet companies still do not have clear mechanisms of action. Doctorate level pharmacologists and basic researchers can educate MSLs on new mechanisms of an older compound, or take research in new, previously unknown areas. The MSL also is a lover of science. Therefore, it would behoove the MSL to call upon not only clinicians performing academic research on patients, but on other scientists as well. It would also be advantageous to understand the pre-clinical aspects of a disease state and the therapeutics used to treat or prevent the disease state.

Patient Advocacy Groups: There are countless patient advocacy groups (mainly not for profit) that seek sponsors for fundraising efforts, board participation, and other volunteers. If you are passionate about a particular therapeutic area, get involved with these groups.

Medical Education Companies: There are national, regional and local medical education companies. They often seek unrestricted educational grant opportunities for educational programs they are developing. Knowing these organizations can provide a wealth of information and opportunity.

Medical Professional Organizations - national, regional and local: Medical associations, professional not for profit organizations, and allied healthcare professional associations also have educational opportunities. Find out what organizations are in your region and at least introduce yourself to the director or president.

Hospital Associations: Hospital networks and associations lie outside of academia in many cases. Do you know their educational needs within your region? Do you know if they provide CE and CME opportunities? If not, it is probably worth an introduction.

Pharmacy Associations: Contact your state, regional and local pharmacy organizations. What educational opportunities they regularly provide to their members is good information to know.

Other Allied Healthcare Professional Associations: Respiratory therapists, radiographers, nurses, physical therapists, and occupational therapists also have professional organizations. Who provides care beyond the physician in your particular therapeutic area? Get to know the other groups in your area that work with patients. A lot can be learned from these healthcare professionals.

Schools of Nursing, Pharmacy, Medicine and Other Allied Healthcare Professions: Currently, there are over 100 schools of pharmacy in the United States. Where are the schools in your region? If you locate them, attempt to call on the CME/CE directors and find out what their educational needs are for their school. Also, you can do them a great service by providing to them your company's process for educational grant requests. This process has become very challenging in the past few years for many CE/CME directors, and therefore, the more information you can provide to the directors at schools, the easier you'll make their jobs.

Directors of CME/CE at institutions: All Universities that provide continuing education have an officer or director or department of CME/CE. They usually hold regular conferences in different therapeutic areas or disease states as well. A wealth of information can be gained by meeting with a CE/CME director at your major academic institutions.

State Government/Medicaid Offices: Last but not least, state government typically has educational initiatives surrounding health issues within a particular state. It would be wise to understand what issues are affecting the states you work in. However, calling on state or government officials may require specific licensure as a lobbyist in some states, so check with your company's policies and/or legal department on how to approach state government officials. Bureaucracy aside, if you have something you can share on how to improve patient care relating to a large initiative within the state, it will be worth the red tape if you can improve the lives of the patients.

Regulatory Compliance

Anyone that has worked in healthcare understands already that it is one of the most heavily regulated industries. Therefore, anyone working in and around healthcare must be cognizant of all the regulatory guidelines applicable to the role, the company, and the industry. There are several acronyms a new MSL should become aware of, understand, and integrate into their work (see Appendix C). In order to better understand the regulatory environment in which the MSL works, read Chris Hall's interview below.

Legal/Regulatory Issues for the MSL To Consider: Interview with Christopher R. Hall, Esq.

Drawing on his experience as a former federal prosecutor, Christopher Hall represents corporations and individuals facing allegations of wrongful conduct. He assists clients with internal and government investigations, enforcement actions, and related proceedings such as False Claims Act and securities fraud proceedings. Mr. Hall places particular emphasis on the heath care, financial service, defense, and government contract industries. His subject matter expertise includes false claims, wire, mail, securities, and tax fraud, government contract fraud, export law violations, and public corruption. He has tried more than 30 cases to verdict in federal courts.

Recent engagements include retention by a global pharmaceutical company in connection with an internal investigation of off-label sales of FDA approved drugs and devices, and an assessment of the company's product lines. Mr. Hall also has been engaged to represent a health care provider in a criminal and civil probe of alleged Anti-Kickback, Stark and False Claims Act violations. Mr. Hall also has been engaged to represent a company in a criminal probe of the public finance markets by the Antitrust Division of the U.S. Department of Justice. Separately, Mr. Hall and other members of the firm's White Collar practice group have been engaged to represent the audit committee of a publicly traded company in connection with an internal investigation of possible Sarbanes-Oxley violations. Finally, Mr. Hall has been engaged to represent the subject of a Foreign Corrupt Practices Act investigation. Mr. Hall's law firm, Saul, Ewing LLP, publishes more extensive information about his experience, publications, and speaking engagements on the firm's website at www.saul.com.

The pharmaceutical industry is one of the most highly regulated industries in the world. Please provide an overview of top guidelines & agencies MSLs need to be aware of and understand.

There are five major sources for guidance regarding interactions between the pharmaceutical industry and healthcare professionals:

1. The Pharmaceutical Research and Manufacturers of America (PhRMA) code on interactions with healthcare professionals - available online at: http://www.phrma.org/files/PhRMA%20Code.pdf.

2. The Advanced Medical Technology Association (AdvaMED) code of ethics on interactions with healthcare professionals (for medical device companies) - available online at: http://www.advamed.org/NR/rdonlyres/FA437A5F-4C75-43B2-A900-C9470BA8DFA7/0/coe_with_faqs_41505.pdf.

3. The Accreditation Council on Continuing Medical Education (ACCME) Standards for Commercial Support^{SM} - available

online at: http://www.accme.org/dir_docs/doc_upload/68b2902a-fb73-44d1-8725-80a1504e520c_uploaddocument.pdf.

4. Two guidance documents from the American Medical Association (AMA):

   a. Ethical guidelines for gifts to physicians from industry 8.061 - available online at: http://www.ama-assn.org/ama/pub/category/5689.html.

   b. Guidelines on gifts to physicians: Opinion E-8.061: clarifying addendum - available online at: http://www.ama-assn.org/ama/upload/mm/369/gifts_clarification.pdf.

5. Finally, MSLs can get a more global view of the compliance landscape from a number of guidance documents published by the Health and Human Services (HHS) Office of the Inspector General (OIG). See: www.oig.hhs.gov. The OIG provides pharmaceutical manufacturers with Compliance Program Guidance, fraud alerts, advisory opinions, and safe harbor regulations. See: http://oig.hhs.gov/authorities/docs/03/050503FRCPGPharmac.pdf.

MSLs work within a gray zone when it comes to off label discussions with thought leaders. What advice would you provide to a new MSL just starting out on how to ensure adherence to the guidelines?

There's a theme that pervades the government regulations and case law regarding the promotion of prescription drugs: fair, balanced, and truthful disclosure. This should be the MSL's guiding principle. MSLs can avoid virtually every promotional pitfall if they disclose sources of financial support and present materials in a fair, balanced, and truthful manner.

This lodestar, moreover, is not at odds with their employers' business objectives. The "gate keepers" to the drug markets—thought leaders and prescribing physicians—comprise a highly sophisticated and intelligent group. The MSL's single most valuable asset in addressing this audience is credibility. A full, honest, fair, and balanced discussion of materials provides physicians with invaluable assistance as they attempt to evaluate the efficacy and safety of a product. Doctors return the favor when it comes time to decide which of a number of competing drugs gets preferred status on a hospital's formulary.

Moreover, reputations, like some diseases, are viral. They spread, good and bad. Doctors, like judges and lawyers, talk amongst themselves. They quickly figure out whom they can rely upon for truthful information.

MSLs of the past often went on joint calls with sales representatives. Should this still occur, and if so, when is it appropriate?

A confluence of legal and business objectives has shifted the industry away from joint calls. By separating the two functions, pharmaceutical companies avoid the appearance of promoting off-label applications when they disseminate truthful scientific information about unapproved uses. Assuming resources are not an issue, companies risk undercutting their off-label educational objectives if they combine the two roles.

When and how is it appropriate for MSLs to talk about off label information?

A MSL can truthfully answer any question posed by a physician. And they should. Their dissemination of scientific knowledge serves the public's interest.

Some of the adverse legal cases regarding off-label promotion stem from so-called "bait" tactics by sales representatives. The term "bait" refers to statements by sales representatives by which they intentionally elicit questions from physicians regarding the off-label use of a drug. They then seek refuge (at their peril) under the first amendment for the off-label conversation that ensues.

While this sort of contrivance presents a risky business and should be avoided, an MSL need never shy away from providing truthful, fair, and balanced information in response to a bona fide question initiated by a physician.

Is it important to document off label conversations?

Yes. Hindsight distorts reality. Government investigators assess whether conduct warrants sanctions by first reconstructing the sequence of events which prompted the inquiry. This reconstruction process, by its very nature, is always incomplete. Nuances and context are lost. And that can lead the government to develop a distorted view about the motivations and objectives of a company or an individual, however well intended they may be.

Good documentation helps to correct the "20-20 hindsight" problem. For example, a government inquiry into the question whether a company unlawfully promoted a drug off-label will, in part, focus on communications between the company and the physician and who, if anyone, initiated the off-label discussion. A note by an MSL reflecting the doctor's inquiry would clarify the issue: "I received a call today from Dr. X regarding the off-label use of drug Y and in response I sent her article Z and made myself available for follow up."

MSLs should of course follow their company's protocol for the capture of contact information, the dissemination of scientific materials, and compliance with copyright laws. MSLs should also draft business records professionally, and avoid careless or informal language, which a third-party might later misconstrue.

Finally, MSL departments should study with care the question of who mines and uses records prepared by MSLs regarding physician contact. Companies should decide how to use MSL records in an intentional and thoughtful manner, all with an eye on how the government might view the matter three years later.

What is the ideal way for a company to document off label requests for information from thought leaders that are asked of MSLs?

There is no magic to it. Classic "work-flow" rules apply. The system should regularly prompt, and require, MSLs to record the information. Periodic assessments should verify that the process works. See the discussion of the FDAMA regulations (now "sunsetted") that follows.

Should MSLs always leave behind a package insert of whatever product(s) they discussed with a thought leader?

The law does not require MSLs to leave behind the package label. That said, it can never hurt to disseminate truthful information. Package inserts are very technical documents, and are FDA-approved. The government would be hard pressed to characterize that practice as promotional. Again, the law does not require the practice, but I don't see any downside from a legal point of view.

MSLs in the past used to be able to share publications with thought leaders. Subsequent issues arose around this practice. Can you share the history and what the ideal situation should be today for MSLs?

A federal district court in Washington D.C., and later the D.C. Circuit Court of Appeals, laid the groundwork for this topic in a case styled *Washington Legal Foundation v. Shalala,* commonly referred to as the *WLF* case. In a nutshell, the Court of Appeals resolved a conflict between the first amendment, on the one hand, and regulations restricting the dissemination of enduring scientific materials, on the other. The regulations, set forth in section 401 of the Food and Drug Administration Modernization Act of 1997 (FDAMA), created a precise set of procedures for the distribution of off-label publications to physicians. The Court balanced the competing Constitutional and regulatory interests by casting the FDAMA regulations as "safe-harbor" provisions. In short, the Court held that the government could not impose sanctions upon a pharmaceutical company if the manufacturer abided by the safe-harbor dissemination regulations. While these FDAMA provisions "sunsetted" by their own terms in 2006, they continue to provide industry guidance.

There has been a change from a regionalized or local model for grant approvals to a centralized model for grant approvals with the MSLs out of the process in some instances. Can MSLs still provide grants to thought leaders and/or institutions for CME and/or investigator initiated trial programs? If so, what do they need to be cognizant of when evaluating a grant opportunity?

The centralized administration of grants permits a pharmaceutical company to efficiently identify needs, ration resources, monitor progress, and evaluate performance. This formula also provides a legal benefit. The government scrutinizes grants to determine whether the grantor truly needed and received the work for which it paid. The government will view a grant as a bribe or a kickback if it did not. Centralized departments have the resources to select worthy grantees and to actively monitor their work. MSLs typically have more pressing concerns.

Finally, the pharmaceutical industry has also been spending billions on research and development of new compounds and not coming up with new product innovations. Can you share how the MSL could foster innovation while at the same time be compliant?

The MSL plays an exciting and important role in the effort to foster innovation. The discovery of new drugs and applications does not take place only in the lab. It depends in large part upon the interplay between researchers and clinicians in the field. The MSL serves as a crucial bridge between these two worlds. MSLs can navigate the complex regulations, which govern their work by adhering to the principles of full disclosure, truthful information, and balanced presentation. That, together with professional adherence to company workflow rules, will ensure a unity of purpose between compliance and growth.

This chapter concludes with perspectives from Dr. Michael Hamann. He shares his expertise in global and international MSL programs.

The International MSL Perspective: Interview with Michael Hamann, PhD

Michael Hamann, PhD Has been working for 12 years in the pharmaceutical industry. He obtained his PhD in Physiology from the Medical Department of Philipps-University in Marburg, Germany. Dr. Hamann started during 1994 in Research in Hamburg, Germany and held various management positions at a large global pharmaceutical company, including Head of Quality Assurance, Project Management, and Head of Oncology Medical Liaisons. He started to build the Oncology Medical Liaison organization for this company in Germany in 2002, as one of their first European Medical Liaison organizations. Since 2005, he has served as the Director of Medical Affairs at one of the largest global biotechnology companies in Europe in Zug, Switzerland. He is currently responsible for regional medical liaison strategies and processes across all therapeutic areas for the international region (Europe, Australia, Russia and Latin America).

How did you get involved in MSL programs internationally?

I became involved in MSL programs in 2002 when I started a local program in Germany for a global pharmaceutical company. Even at that time we were exchanging views and experiences across Europe. It became clear to me that continual scientific discussions and engagement of physicians in research programs was essential to evolve medical practice and help improve patients' lives and optimize their therapies. This was especially true for scientifically driven therapeutic areas such as oncology.

Where do MSL programs exist outside the US?

MSL programs exist in most of the western European countries and Australia. Increasingly they are present in Eastern Europe, and are arriving in Russia.

My feeling is that the regions with MSL programs are following the global expansion we saw with clinical trials. It will be interesting to see when the first programs will start in India, for example.

Do outside US (OUS) companies have pharmaceutical representatives, and if so, do the same differences exist between MSLs and representatives as in the US?

They all have pharmaceutical representatives. MSLs are clearly differentiated and are quite comparable to the US.

What are 3-5 major differences between MSLs in the US and outside the US?

In the US, the MSL role is very well defined. Outside the US, however, the role description might vary by country. One factor is the structure and size of a country. Particularly in smaller countries, the role can be broader, covering a range of activities in a medical department of a company. Subsequently, these MSLs might cover multiple therapeutic areas and more products. In terms of qualification, we have many PhDs and a significant number of MDs.

If a US based company wants to move into the European market or into OUS MSL programs, what things should a MSL director think about in order to initiate a program?

It is most important to acknowledge cultural differences. You need to understand what level of scientific communication already exists and the structures of the scientific community, otherwise a well-designed MSL program might fail.

How can MSL teams be structured and organized outside the US?

There are three different models. A centralized model with corporate funding and direction, a mixed model with corporate funding but local direction, or a decentralized approach with local funding and direction. Which model you finally choose will depend on various factors such as therapeutic area, countries involved, and company culture.

Can US MSLs be 'transplanted' into other countries or has this not been successful?

I am not aware of any such successful 'transplantation'. It would certainly require substantial training and understanding of the local situation. From that point of view I would always prefer local deployment.

What issues are unique to the MSL working outside the US?

Despite common regulations across countries, the more detailed issues are always country specific. An MSL director for a non-US program might spend a significant proportion of work on understanding local regulations and differences. For example, countries that have recently joined the EU are constantly changing regulations in adapting to common EU regulations. This can lead to significant delays in research programs. You also have to think about travel issues, especially if visas are needed for some MSLs. An *ad hoc* training or meeting is not possible under these circumstances.

Chapter 5:
Customers' Viewpoints

As previously stated, the MSL serves a variety of thought leaders. This chapter contains interviews with four different customers called upon by field based MSLs. The four thought leaders include: Dr. Chris Bojrab, Dr. Ronald Chervin, Dr. James Simon, and Dr. Paul Keck. Each of these customers has different needs and requires different things from the field based MSL. Similar questions were asked of each to see the variety of their responses, based upon their individual practice and professional setting.

Interview with Chris Bojrab, MD

Dr. Bojrab is a psychiatrist, psychopharmacologist, and president of a large multidisciplinary private behavioral health practice in the Midwest. He also sees patients in a number of other practice settings including an outpatient clinic with intensive outpatient and partial hospitalization services for adolescents and adults, a child/adolescent residential setting, and in the nursing home setting. He is a national lecturer regarding the pharmacologic treatment of a variety of psychiatric illnesses and other medical conditions that impact mood, anxiety, energy, cognition, chronic pain, and sleep.

Can you share your experiences as a speaker and clinician regarding the differences between pharmaceutical representatives and MSLs?

I have more sales representatives than MSLs calling on me. But they bring very different things to my office because of the differences in their training. The information the representatives bring is useful for specific product training, but many lack a science background. The representatives know the nuts and bolts of the product, but the MSL can discuss the clinical implications of a drug – metabolism, kinetics, pharmacology, etc. The MSLs know pharmacology in some cases better than physicians; thus, there is a more collegial feeling to the conversation between a physician and a MSL. The MSLs are more holistic in their knowledge of other medications and are more familiar with the literature.

What other things has a MSL brought to you besides technical educational knowledge around a product or a disease state?

The MSLs used to provide a direct conduit for grants to support medical education, but most companies now provide a centralized repository for grants. MSLs were an integral part in the past to the grant process, but that is no longer the case.

You are also a speaker for several companies. What has the MSL provided in terms of speaker training?

The MSLs have more up to date information – they are more in touch with research and development and are more aware of studies planned, studies currently ongoing, biostatistics, and recently published trial information. We used to participate in clinical research, but after a colleague that was the principal investigator left, we stopped. Both of the trial opportunities came through internal research teams (clinical operations and via a mass mailing opportunity). I have spoken to some MSLs regarding research opportunities, but it is trending downward and I have not spoken to as many MSLs regarding participation in research projects recently. It strikes me that a MSL could bring significant value to an organization by being involved in earlier stage research opportunities. As a busy clinical practice, clinical sites would be ideal for many clinical phase II/III studies because there are a higher number of patients. Electronic medical records in busy clinical practices would also be a huge advantage to speed up clinical research. Outcomes would be easier to obtain, particularly for investigator-initiated trials. More post-hoc analyses could be completed.

Traditionally, MSLs enjoyed more freedom and flexibility in what they could discuss, and less formality with "off label" uses. Not that MSLs actively promote or discuss off label, but there is a collegial discussion rather than a formal, rote response. The quality of information is richer from a MSL. I would like to see more of a bridge to basic science people within a drug company. The companies I have had the opportunity to tour internally are interesting, because one can understand why things happen to a compound in the way that they are brought to market, and why medicine costs a lot. There may be an opportunity for MSLs to be more of a conduit or liaison between pre-clinical research and the clinician. Perhaps the MSL could even bring these two groups together.

Do you see a difference among MSLs in their professional backgrounds and the value that they bring to you?

I think there are differences. The MD MSL has a higher level of trust and respect because they have worked in clinical practice and know what it is like. PharmDs are closer to the clinical aspect than the PhD researcher. The PhDs can often bring better science in terms of preclinical and bench information. But I never had a PhD MSL that didn't know what they were talking about – they just didn't go near the clinical conversation. But within the frames of science and knowledge base, I have not seen many differences. Most people were respectful and aware of the differences and didn't pretend to be something they weren't trained to be.

The ideal MSL team is probably a mix of all professional backgrounds.

Yes. And it is product dependent. If you have a product that is a 'me too'—another in a line of similar products with the same mechanism of action (MOA) – a PhD wouldn't be as valuable to the compound as a clinician that has experience with the product. Whereas if you have a product with a novel or different mechanism of action from everything else on the market, the value of a PhD may increase due to the value of the science surrounding the compound. Case in point: I met with a MSL a couple of weeks ago, she clearly had a love of pharmacology and had the ability to provide me with some pre-clinical publication data that perhaps many MDs or PharmDs might not have at the top of their mind. It is science for science's sake.

What else would you want to see from MSLs?

I would like to just see more of them, period. My interactions with them are limited. If you asked the majority of physicians, they would not know the differences between MSLs and pharmaceutical representatives. Physicians assume that MSLs are representatives that have been working longer. I do a lot of consulting, speaking, and advisory board work with drug companies, so I would like to see as many MSLs as possible in psychiatry, but there are only a few that I see quarterly or so per year. One of the companies I do a lot of speaking for has a MSL that I've contacted a couple of times about some questions that came up, but I never hear from that individual, which is rather disturbing.

Do you think the MSL position will endure?

I think so, and I think it will take on an even more important role with the cuts in sales forces now. Sales forces are bloated now. I don't see a lot of duplication in representatives calling on my office, but many of my general practice friends are seeing multiple representatives with the same compounds. The typical sales call now costs $200-$600 per call, and the return on investment is no longer making sense. The sales model will go back to being more of a 'detail person' and there may be more opportunity for the MSL to morph more into a sales role, because physicians are going to demand more technical knowledge and value to the clinician with less representatives around.

Interview with Ron Chervin, MD, MS

Ronald D. Chervin, MD, MS, is Professor of Neurology and holds the Michael S. Aldrich Collegiate Professorship in Sleep Medicine at the University of Michigan in Ann Arbor. He directs the UM Sleep Disorders Center, which maintains two sleep laboratories, innovative multidisciplinary sleep clinics, competitive clinical and research fellowships, and productive research programs. Dr. Chervin's research has focused on sleep-disordered breathing, polysomnographic methods, assessment of sleepiness, and neurobehavioral consequences of sleep disorders in children. He serves on the Board of Directors of the Sleep Research Society, and on the editorial boards of Sleep, Sleep Medicine, and the Journal of Clinical Sleep Medicine. In this interview, Dr. Chervin shares his personal opinions regarding the MSL, they are not the opinions of the academic institution in which he is employed.

How do you view the differences between MSLs and pharmaceutical sales representatives?

In my experience, there are wide differences. Sales people have a narrow focus. Sales representatives may come from a background related to the product, and are generally well educated about the product, but they are limited in what they can say by what the company and the FDA have approved for them to say. It is a world of difference with an MSL. When I meet with an MSL, I feel like I am talking to a scientist who is not under the same restrictions. The MSL generally has a broader background and a wider ranging scope regarding therapeutics, which leads to a more satisfying conversation. I have had informative experiences with sales representatives, but the conversation tends to end in 'let me send you more information', whereas the MSL can more readily tell me the limit of existing knowledge. MSLs have more interest in the research within the field and have been helpful in other ways. They have often been supportive for educational aims of our center. They have brought in interesting speakers and helped us to put on CME programs. They have also assisted in finding ways to fund research. The MSLs tend to know more about investigator-initiated research opportunities than sales representatives. Many companies don't offer us any contact with MSLs. But in most cases, the experiences with MSLs I have met with have been very rewarding. I have never turned down an opportunity to meet with an MSL.

What beyond technical information have MSLs brought to you, your department, and your university?

The MSLs have been more forthcoming with pipeline information and what is under investigation. I often get an 'I don't know' from a sales representative. The MSLs are more willing to say what they know about comparing and contrasting competitive products, or the range of products within a therapeutic area. Obviously as a representative of a company, at the end of the day, the MSL still works for the pharmaceutical company, but the MSL doesn't have the pressure of being a sales person. They have more objective and broader views at times. The MSLs can also bring resources, like funding and other organizational resources. The salespeople tend to bring pens and books, and there are many restrictions now as to how companies can support medicine. But the MSLs can really help with programs. Programs in academic centers are unfortunately not always well supported, especially when it comes to discretionary funding. At times, there can be a real shortage. In a specific example, through working with industry contacts and MSLs we have been able to support an annual sleep CME course. They also have helped to fund a research endowment. We also have help every year to sponsor a lectureship in sleep medicine, which was actually arranged through the help of a sales representative. My impression is that the MSL, in comparison to a sales representative, generally has access to a higher level of corporate resources.

MSLs have very different backgrounds (PharmD, PhD, MD or other healthcare professions). Do you see a difference among the MSLs that work with you in terms of their value to you?

I've only had experience working with PharmDs and PhDs. The differences are quite clear in sales versus MSLs. Sales are reasonably well educated on medication being discussed. The PharmD or PhD offers the chance to talk to another professional. I'm not saying the representatives aren't useful, but in the academic setting it is more useful to talk to the doctorate level person. I have not seen a difference among MSLs in terms of professional background, but I have not worked with a large number of MSLs.

What frustrates you about working with the pharmaceutical industry?

I have not had a lot of frustrations working with the pharmaceutical industry. Time is always a scramble. If sales people come too often, that is a problem, but I have not personally experienced this as an issue. There has been public concern with the potential influence of the pharmaceutical industry on prescribing patterns, particularly among trainees. Although I am aware of this issue, I have not personally been aware of abuse within the system. Sometimes I'm frustrated that a salesperson cannot always answer a question. Generally, most are good about getting me a follow-up packet of information on the topic. It may be two weeks later and the representatives are generally good about the follow up - I can't complain in that regard. I'm aware of issues with the FDA and drug safety, and I always maintain a level of skepticism based upon the source. Sometimes when I work with a sales representative I do not always get data, or the level of data that I had hoped for. But in general, my relationship has been positive. I do not typically speak for companies or perform a lot of pharmaceutical company sponsored research. I generally have little to disclose when giving a talk or writing a publication.

How is working with the pharmaceutical industry rewarding?

It is rewarding in resource availability - we would rarely have that, otherwise, in an academic center. Also, working with the industry has been good for brainstorming research ideas and collaboration. Those interactions can be interesting and educational. I can sometimes provide better care or am better able to educate patients based upon MSL interactions. I learn about precautions. For example, I became aware of dopamine agonists and heart failure. I was not aware of this until someone sponsoring a journal club told us about it. I quickly had an opportunity to make use of this information in my clinical practice. The industry can help us keep up with the literature; it is impossible to read every paper in every journal every week or month. When we see sales representatives for medications we are commonly using, it is more targeted and better than the average ad in the average journal.

What else would you ideally want to see from MSLs that they might not be able to bring to you currently?

Funding, we could always use more. I have also occasionally asked MSLs if they are able to provide research data from large trials, most often sponsored by the pharmaceutical companies, for retrospective analysis. I have not aggressively pursued this, nor have I been successful with the concept. If companies would provide more liberal access for retrospective data analysis, research interests would be advanced. MSLs have generally have been very helpful. The ones I have met have been the most helpful industry representatives that I have seen, because they are interested in what we are doing and they want to know how they can help. What more could you ask for?

I also have to say that it does help when we have a long-standing relationship with industry representatives. You know what to expect from them, and you know whom to contact when you have questions. It appears that pharmaceutical sales people rotate frequently or get promoted, but this also seems to occur with the MSL at a slower rate. Obviously people need to be promoted, and many sales people utilize the sales representative position as an entry point into the industry. Not as true of MSLs. The typical tenure for a MSL seems to be about 2 years. It would be ideal if there was some way MSLs could be promoted but still maintain their original contacts. There is value in having a person be in a consistent role.

Do you think the MSL position will endure?

I honestly have no way of knowing. I was not aware of what an MSL was 6 years ago. I've been on the faculty here for 13 years, and for half of my time here I did not work with MSLs. I didn't know what one was or whether we had one from various companies. I am not sure if the MSL position is valuable in clinical practice settings, but I definitely see value for them in the academic setting.

I am not clear as to whom MSLs have contact with, within the company, but I would hope they have frequent conversations with high-level management within their respective companies. The MSLs are the only people in the pharmaceutical company out in the field working as scientists with other scientists. If a company truly wants to have their finger on the pulse and know what the academic community is thinking, the executives would be well served to have close contact with the MSLs, just as the MSLs are having close contact with academic faculty.

Interview with James A. Simon, MD, CCD, FACOG

Dr. Simon is Clinical Professor of Obstetrics and Gynecology at the George Washington University School of Medicine in Washington, DC. Dr. Simon's research has been supported by more than 120 research grants, contracts and scholarships from a wide range of sponsors (NIH, AHA, pharmaceutical industry, etc.). He is a past president of the North American Menopause Society (NAMS) and of the Washington Gynecological Society, and currently sits on the Board of Directors of the International Society for Clinical Densitometry (ISCD). Dr. Simon has been selected for "Top Washington Physicians", "America's Top Obstetricians and Gynecologists", and "The Best Doctors in America". He is an author or co-author on more than 160 articles, chapters, and proceedings, including several prize-winning papers, and the paperback book: Restore Yourself: A Woman's Guide to Reviving Her Sexual Desire and Passion for Life.

Can you share your experiences as a thought leader - the differences you perceive between pharmaceutical representatives and MSLs?

There is no comparison. MSLs in my experience are scientists or PharmDs and have clinical experience and knowledge that is far beyond the sales representative. They have a different focus and perspective. If a MSL is trying to sell me something, I get rid of them. I see MSLs as professionals, as someone of similar interests if not similar backgrounds. They are working to disseminate peer reviewed information, formulate new questions, develop new science, and to encourage me to submit investigator-initiated research proposals, results of my current research, and other information back to their respective organizations.

What value has the MSL brought to you besides technical educational knowledge and materials around a product or disease state?

I think the good MSLs are particularly focused on a small area of scientific inquiry and they bring me literature and biographical information I might not normally find. For example, some bring abstracts on literature that might be published in other languages, or peer reviewed articles from journals I don't normally read. Sometimes they are focused on their own products or obscure data that does not always get published in peer reviewed journals. They bring me information that I would not normally find on my own.

What about research?

I would say that the MSLs I know well act as a sounding board for me as they know the therapeutic areas in which I work. They may also have insights into the company and know what the company is interested in funding, and help me understand how to optimally share a research idea with their company. For example, if I submitted a research proposal on my own to the company, I might not have the perspective I need to optimize the ability of the research to be funded.

Do you see a difference among MSLs in their professional backgrounds and the value that they bring to you (PhDs versus PharmDs versus MDs)?

I think it depends on what you are looking for - I think a good PhD has a very good feel for the clinical. A really good PharmD may not know the science from the "hands-on" position, but may be able to research the literature and come to know the field from that perspective. I don't see a big difference between them. The question should be more about how willing MSLs are to work and increase their personal strengths and overcome any specific weaknesses. It is individualized. It is important to bring one's strengths to the job. I have some MSLs that do not have advanced degrees, and they are spectacular. They aren't resting on their academic laurels (or lack thereof); they are going out and aggressively getting new ideas and keeping up on the research. Learning is new and exciting to them too.

What is your definition of a good MSL?

A good MSL brings me some value. Value is not the same as money. Value can be general information or research opportunities for the whole field, it could be general information about the MSL's particular product as it relates to the field, or general and specific information for or against conventional wisdom. But they always bring me something. The people who are not good MSLs are focused more on the social aspects of their job. They feed me well, but they do not provide information, research funding, or education. They can sustain me many times over with a well-prepared slide set that reflects what is new in the scientific literature even if it is not favorable to their product. This helps me because it saves me the time of constructing my own slides.

What frustrates you about working with the pharmaceutical industry (if anything)?

My feelings about the pharmaceutical industry are generally positive. Occasionally, I get a new sales representative that doesn't know his or her information. They assume their knowledge level is at my level. They don't know what they are doing when they present one of the papers I originally authored or go through a sales aid that has data from a publication that I published and don't know it is mine. This has happened to me several times over the course of my career, and when it does, I simply provide to them a signed copy of the article and let them know that they can return to my office, provide samples, and discuss data with me only when they are ready.

What is the most rewarding thing about working with MSLs?

I think they can be a very good liaison between me and my colleagues in the same city and region and help me to be a more interactive and participatory clinician and scientist. I realize this isn't in their job description, but it is important. They are the conduits to efficient networking within a particular field of endeavor. Recognizing that in a small group, interactions are always on the one hand competitive, and on the other hand collaborative, they can cut through and streamline communications by making people aware of meetings, or visiting professors that might not come across my personal radar. For example, the world authority on systemic mastocytosis is coming to Washington DC to talk about this disorder. Let's say I wanted to know something specific about systemic mastocytosis, possibly bone loss or osteoporosis. Maybe I will go to grand rounds on the morning of this speaker just to learn about this - but if this MSL didn't tell me about sponsoring this event, I would have never heard about it. Or, the MSL might be bringing a really noteworthy person to town for grand rounds at a university, but I can't go. However, they may also be coming to town for a round table or a promotional talk. At a minimum, I would have the opportunity to meet them, interact with them and hear what they have to say. Most MSLs don't do promotional programs, but in many cases they do case reviews or journal clubs, which are also of benefit. It is extremely rewarding to have a world authority come to town because the level of conversation is way above the norm.

What additionally would you like to see MSLs bring to you that they do not currently?

I think the things I'm getting from my MSLs that I've always received, it is just that there is a lot more red tape. The MSLs currently are part of medical affairs at many companies. Even though they are part of medical affairs, any projects or research ideas still need to go through the same funding channels and process as someone from the sales and marketing side - the regulatory and legal channels are often the same within a company. For example, I would sit down with a MSL and go through a new literature slide set, and the MSL asks if I would like a copy, but he can't email me a copy. Why can't he email me? He has to go back to the company, complete the required the paper work, go thru legal and mystical third parties, and then and only then, will I finally receive the slides.

In terms of investigator-initiated projects, it used to be that the MSL would work directly with the investigator to develop the proposal and then throw it "in the hopper" of other proposals. Now, everything is done centrally. Now the MSL is just trying to figure out who is in charge of investigator-initiated research internally, what they are interested in from a research perspective, and trying to understand the operational aspects of submitting the proposal. Research development has been taken away from them, centralized, bureaucratized, and depersonalized.

Do you think the MSL position will endure?

No, because in our current business model, documenting return on investment (ROI) for the MSL has and continues to be extra difficult. As finances shrink and consolidation occurs, new drugs don't get approved, and costs of drug development go up, the role of the MSL becomes inordinately expensive when you can't prove ROI. I have the opportunity to talk to many managers of MSLs, and the common refrain is, "I don't know how I can do this job with 6 MSLs for the whole country when they are on planes 6 days per week". Or, "I only have enough people so they can see their thought leaders once in a blue moon". If the MSLs are not local, or at the worst regional, they miss out on the important parochial events and decrease their value - they can't be the local liaison if they are simply not here enough.

Interview with Paul Keck, MD

Paul E. Keck, Jr., MD, is the Craig and Frances Lindner Professor of Psychiatry and Neuroscience and Executive Vice Chairman of the Department of Psychiatry at the University of Cincinnati College of Medicine. He is also President-CEO of the Lindner Center of HOPE, a state-of-the-science, UC-affiliated comprehensive mental health center in Mason, Ohio. Dr. Keck has conducted research in bipolar disorder and psychopharmacology.

A magna cum laude and Phi Beta Kappa graduate of Dartmouth College, Dr. Keck received his MD with honors from the Mount Sinai School of Medicine, New York, NY. He served his internship in Internal Medicine at the Beth Israel Medical Center in New York and completed his residency training in Psychiatry at McLean Hospital, Belmont, MA. Dr. Keck remained on faculty at McLean and Harvard Medical School before joining the Department of Psychiatry at the University of Cincinnati in 1991.

Dr. Keck is the author of over 300 scientific papers in leading medical journals and was the 8th most cited scientist in the world published in the fields of psychiatry and psychology over the past decade. He has also contributed over 160 reviews and chapters to major psychiatric textbooks. Dr. Keck is the editor or author of 6 scientific books and serves on the editorial boards of 7 journals. He also serves on the American Psychiatric Association's Workgroup to Develop Practice Guidelines for Treatment of Patients with Bipolar Disorders and currently serves on the APA Institute for Research and Education. Dr. Keck was a member of the FDA Psychopharmacologic Drug Advisory Committee.

Dr. Keck is the recipient of numerous honors, including the Gerald Klerman Young Investigator Award from the National Depressive and Manic-Depressive Association (NDMDA); the Gerald Klerman Senior Investigator Award from the Depression & Bipolar Support Alliance (DBSA); the Exemplary Psychiatrist Award from the National Alliance of the Mentally Ill (NAMI); the Philip Isenberg Teaching Award from Harvard Medical School; the Nancy C A Roeske Certificate for medical student education from the American Psychiatric Association; Distinguished Fellow of the American Psychiatric Association; the Wyeth-Ayerst AADPRT Mentorship Award; two Communicator Awards for Continuing Medical Education; the Outstanding Physician Partner Award of the Postgraduate Institute for Medicine; and two Golden Apple Teaching Awards from the University of Cincinnati College of Medicine. He is listed as one of the Best Doctors in Cincinnati by Cincinnati Magazine; The Best Doctors in America, a directory of the top one percent of physicians in the United States as rated by their peers; and as one of the nation's Best Mental Health Experts by Good Housekeeping Magazine.

How do you view the differences between MSLs and pharmaceutical sales representatives?

In my view, pharmaceutical sales representatives provide on-label information about their product, and to assist healthcare providers in prescribing appropriately and safely. MSLs have a largely independent and separate role. I think the MSL role has evolved and differs widely by company. In the broad sense, MSLs provide new information that is more generally related to neuroscience. I have been given information on drug development in order to solicit feedback on the development plan or possible innovative ideas regarding new applications for emerging or already marketed products.

What beyond technical information have MSLs brought to you, your department, and your university?

I can't speak to the university, but MSLs have brought mainly information. Specifically, they have brought new research findings that may not be in the form of articles yet, but have been presented at scientific meetings. Information they provide is non-proprietary and in the public domain, but ahead of wide dissemination. Also, MSLs shared information about new therapeutic areas the company is interested in. Lastly, they have served as a conduit for investigator initiated research studies and as the champions for unrestricted educational grants.

MSLs have very different backgrounds (PharmD, PhD, MD or other allied healthcare professions). Do you see a difference among the MSLs that work with you in terms of their value to you?

I have not seen a huge difference in the quality of MSL performance based upon their professional credentials. Knowledge of the field (in my case psychiatry) has been the essential feature. Knowledge of psychopharmacology is also important. The intangibles, such as communication skills, follow through, and passion for the job, have been the things that have distinguished MSLs from being good or not so good.

What frustrates you about working with the pharmaceutical industry, if anything?

Lack of, or inconsistent communication (which includes things like multiple points of contact at one company) is frustrating. Also, lack of follow through by companies is somewhat frustrating.

How is working with the pharmaceutical industry rewarding?

I think the main, or most rewarding areas, have been in contributing to the treatment advances that have been so substantial for people with mental illnesses. Working with the industry is the quickest way to achieve these advances in medicine. Also, to explore the application of approved medications in new therapeutic areas is greatly rewarding and has the potential to improve patient care.

What else would you ideally want to see from MSLs that they might not be able to bring to you currently?

I think the industry has been ambivalent about the role of the MSL, and they have been underutilized. I think the more clearly defined the role of the MSL, the better. Furthermore, it would be helpful to know at the beginning whether or not the MSL has any budgetary or advisory role with the parent company. It is critical for MSLs to articulate what is and is not within their scope at the beginning of the relationship.

Do you think the MSL position will endure?

I am not sure. In the current regulatory climate, which has now shifted to being extremely conservative and restrictive, I think the role of the MSL has become more difficult to carry out. I think the future of the success of MSL roles will depend upon being explicitly clear about what MSLs can or cannot do. I am aware of concerns that part of the MSL's role could be confused as trying to stimulate studies in order to increase off label use of compounds. However, treatment advances often occur when enlightened investigators see a potential application of a product in areas where, for budgetary reasons or other restrictions, the company hasn't wanted to commit to studying.

Part of the frustrations I have had expressed to me by MSLs is that they work at too low of a level in the organization and that their impact and their feedback is either unheard or diluted. When we consider the professional background and credentials of the MSLs, this is fairly surprising. They bring scientific expertise and the ability to evaluate the feedback they receive from the field. I think the present role of MSLs in most organizations is too marginal. The position should either be eliminated or taken to a higher level and given a more prominent role in the organization. Right now it's the worst of both worlds. I think the better way for companies to utilize their MSL teams is to provide a higher-level role in the overall management of the product or the specialty area. For example, have the MSLs working within the neuroscience or oncology division, rather than working on a specific product. MSLs should have access and regular dialogue with executive level management.

Chapter 6:
Research & Technology

Technology is an integral part of the MSL's work. However, in the age of hand held devices, cell phones, and travel, it is a big challenge for the MSL to not only to keep up on new information, but also to manage potential information overload. There are three interviews relevant to the MSL contained in this chapter. First, Lou Ann Fare, a current MSL, discusses general information management in her work. Secondly, Dr. Amy Peak discusses information management relevant to the MSL and drug information. Finally, Michael Taylor of Akuta Corporation discusses MSL metrics and how they relate to call management systems.

Resources For the MSL:
An Interview with LouAnn Fare, MS

LouAnn began her pharmaceutical industry career in 1987 with a marketing internship in Radnor, PA. Soon after, she received her BS in pharmacology/ toxicology from Philadelphia College of Pharmacy and Science. LouAnn has performed biodistribution studies on anti-neoplastic and anti-fungal compounds and monitored clinical studies in pulmonology and cardiology. While working as a Senior Medical Science Liaison for the past nine years, LouAnn has studied various disease states within gastroenterology, cardiology, and neuroscience and has received her MS in Regulatory Affairs/Quality Assurance from Temple University in Philadelphia. In 2005, LouAnn completed a certificate in Competitive Intelligence from Drexel University.

Beyond the traditional medical and publication databases, what types of technology do you utilize during your work as a MSL?

Our team relies heavily on our in-house research librarians to provide us with current key word searches. We also use Thompson Scientific's message mapping system. The subscription resources have proven invaluable. For instance, LexisNexis® and Ovid® are timely and easily searchable. The team also uses a customer relationship management tool. This allows MSLs to understand what paradigms thought leaders across the country are discussing and teaching.

What online resources do you utilize as a MSL, beyond the databases for medical information?

I work in neuroscience, so I tend to go to a lot of neuroscience forums online, such as schizophreniaforum.org, The National Alliance on Mental Illness (www.nami.org), The Pharmaceutical Researchers and Manufacturers of America (www.PhRMA.org), and other sites such as Clinicaltrials.gov. I also visit many conference websites, like psych.org (APA's website), the New Clinical Drug Evaluation Unit (NCDEU, located online at: http://www.nimh.nih.gov/ncdeu/index.cfm), anything that relates to a conference.

How can technology help the MSL work more effectively?

In the past 5 years, major conferences have come really far with online scheduling. With pre-conference access to programs and abstracts, MSLs can organize their time efficiently and effectively before they even arrive at the conference. I determine which scientific sessions would provide the most up-to-date information and which presenters are most likely to deliver a sound scientific and medical presentation. Another plus to having pre-conference access is that, as a team, we can pre-determine when major updates will be released. We have a very good idea of what to seek out based on the previous year's conference information. We are constantly asking ourselves "What is new?", "What updates are expected?", "Was a drug in phase II or III discontinued?". Professional organizations and advocacy groups are also utilizing the Internet to display their mission statements, upcoming events, and membership details. Information is more readily available, particularly for the professional associations, than it was 5 years ago.

When I first started, the palm pilot was a huge boon because you could take your schedule with you. I use a Treo™ now that utilizes the Palm® platform. It helps with maintaining e-mail and communication with work. For example, I can view slide presentations prior to my appointments. External pen drives have also been helpful. If I can provide information without carrying a heavy laptop, I'll use an external pen drive.

How does technology assist the MSL from a legal/regulatory standpoint?

The Office of the Inspector General (OIG) Compliance Program Guidance for Pharmaceutical Manufacturers, FDA.gov, phrma.org, and accme.org are regularly accessed online. Technology can help adopt and formalize best practices on conduct of activity and compliance as a MSL. In fact, I know of a chief compliance officer at a company that was a former MSL. So there are definitely strong ties between regulatory and the MSL world, and technology facilitates that relationship.

What advice would you have for someone interested in becoming a MSL?

I would recommend they obtain an advanced degree. I would also recommend that his or her first position be within a therapeutic area that he or she is interested in. What's key is to not limit yourself to just that one therapeutic area. As always, it will remain important to keep learning and furthering your education in clinical and scientific areas of expertise. Also, it is critical to maintain a fair and balanced method for delivering your information. Good debating skills are also required. Be prepared when meeting with thought leaders. Do your homework and have an idea of the thought leader's scientific interests and major publications before you meet with him or her.

What is the best thing about being a MSL?

As a MSL, you are an agent for change. Being a liaison, you are the bridge between what occurs in clinical practice and how industry interprets clinical practice. Another good thing is that I deal with some of the brightest minds in the world, some who teach, some who maintain clinical practices, and some who do both. Our clinical exchanges may vary depending on the type of healthcare professional I'm talking to. But, by understanding their needs and issues, by knowing historical and current teaching paradigms, and by seeing where future paradigms are trending, I can provide my company with ideas about what kinds of clinical trials should be proposed, what's working in clinical practice and what isn't, and where future investments lie. Most importantly, I think a MSL can be an advocate for researchers, physicians who benefit from the research, and ultimately for the patient. Physicians and patients need to be exposed and have access to unbiased, evidence-based, current information in order to make proper clinical decisions. The MSL is the person who knows and understands the historical evidence as well as the current research and makes use of that knowledge for the betterment of clinical practice.

What is the most challenging thing about being a MSL?

The MSL must have the ability to synthesize and organize a lot of readily retrievable information. While it's true the MSLs must keep abreast of current clinical practice, the most challenging part of this position is determining which studies and trials are unbiased, evidence-based, and well-controlled. Critical analysis of trials and studies is essential. Knowledge management is a large component of the job. The Internet has inundated medical professionals with information. In general, thirty percent more information is in the public domain now versus 20 years ago. The most challenging part of being an MSL is having to sift through all of the information and verify the conclusions based on the way the studies were conducted.

What advice would you give to a MSL just starting out?

First, understand what type of company you work within - small, medium or large - and then determine how your company defines your role as a medical science liaison. Understand where each product's lifecycles are within the therapeutic areas. The product's lifecycle phase will most likely determine how you will conduct yourself in interactions with healthcare professionals.

Embrace the diversity of the group you work within. Everyone has different professional backgrounds, and in those diverse backgrounds lie unique strengths for each team member. Also, MSLs need to focus on the unmet needs of the therapeutic area(s) so that they can ask thought provoking questions of their thought leaders.

This position can be very demanding and isolated. Working from home allows for some freedom; but, because you're working so hard, you may also develop an inclination to forego social invitations. It's important to understand your interpersonal boundaries, to build a strong social network, and to maintain other interests outside of work. During my first year, I was working a lot of hours per day and it felt like I never left the office. But, you have to say to yourself that you are only going to work a set number of hours and then walk away from it.

Keep in mind that with this position, traveling to conferences and other business-related events is a necessity. After these weeklong trips, you may not want to socialize. Maintaining an open dialogue with friends and family is essential in balancing your work-life approach.

Erin Albert and Cathleen Sass

Managing Medical Information: An Interview with Dr. Amy Peak

Dr. Amy Peak is a clinical pharmacist, assistant professor, and Director of Drug Information Services at Butler University. She received her Doctor of Pharmacy degree from Butler University and completed a pharmacy practice residency at St. Vincent Hospitals and Health Services, then was visiting scientist (fellow) at Eli Lilly and Company. Following her formal training, she was Director of Drug Information Services at St. Vincent Hospital and Health Services, prior to going to Butler. She is a past president of the Indiana College of Clinical Pharmacy as well as the founder and past president of the Drug Information Practice and Research Network within the American College of Clinical Pharmacy. Her primary practice, research, and interest areas are related to drug information, information technology, herbal medicine, and medication safety.

As a medical/healthcare information professional, what are your top 3 most utilized websites?

My most utilized websites are definitely subscription based medical information databases. There is so much mis-information on the Internet, it's really important to get your information from a reliable source that has other healthcare professionals creating the content. Typically those services are not provided for free. Even if it's a popular "free" site, it's provided to you by advertising dollars, creating an unavoidable potential for conflicts of interest. I do, however, use a lot of the websites of professional medical organizations as well as government sites such as the National Institutes of Health (NIH), The Centers for Disease Control and Prevention (CDC), and The Food & Drug Administration (FDA) sites (although they are not very user-friendly).

How do you stay up on the literature?

There is so much information out there, it's really essential to create and utilize an effective plan for staying current. Here are some suggestions for creating an effective personal plan for keeping current:

- Manage time effectively. No one has time to read all the new information.

- Realize that more isn't always better. It is so easy to get into the information overload trap.

- Identify one to three resources or services that most efficiently meet your needs and actually commit to reading those regularly. Too many people have great intentions and sign up for lots table of content emails, review services, list-servs, etc. Then they end up deleting them from their in-box without ever reading them. So they sign up for tons and end up reading very little of it.

- Are you a generalist or a specialist? Focus your efforts on those resources that most closely match your personal practice area. What works for a cardiovascular clinical pharmacist practicing in a hospital is going to be very different than what works for regulatory affairs pharmacist practicing in pharmaceutical industry or a community pharmacist practicing at your neighborhood pharmacy.

- Consider using complementary resources, with one focusing on new primary literature and the other focusing more on day-to-day practice. For example, the two things that I can honestly say I read regularly are JournalWatch® and The Pharmacist's Letter®. For me, JournalWatch® is a great way to keep abreast of the newly published studies because their editors search the most common medical journals, find the most pertinent articles, summarize them in about 3 paragraphs, and compile them in an quick, easy to read newsletter that comes out twice monthly. This way I know a little about a lot of the new things that are going on and can then decide if it's worth my time to pull and review the entire study. Pharmacist Letter® fills a completely different niche. It's not focused on primary literature at all. It's focused on what new prescription and over the counter medications are coming out, what new guidelines have been released, changes in pharmacy or medical law, etc. For me, these two resources complement each other really well and are very effective ways of keeping up with new therapies and management strategies without investing an unreasonable amount and effort.

When it comes to literature searches, can you provide an overview of the top 3-5 databases and what the advantages and disadvantages are to each?

Although this seems like a straightforward, easy question, it's actually rather complex, especially because databases change and merge regularly. Additionally many venders, such as OVID®, EBSCOhost®, etc. combine individual databases for a more robust system.

PubMed/Medline® is what most people consider the gold standard. It's the largest primary literature-indexing database for medical literature. It currently indexes about 5000 journals and contains over 17 million citations from as far back as 1950. Although most consider PubMed the gold standard, it is getting so big that it can be problematic. Its size is a blessing and a curse. It is common to get literally thousands of hits for a given search. So to use PubMed efficiently, you need be to very good at limiting your search strategies and using tools such as medical subject heading (MeSH) terms, etc. In addition to being the biggest database, it's free. Everyone has access to PubMed. However, only a small fraction of the citations are available full text format if the free version is used. Thus most institutions purchase a database from a vender that extracts the citation data from PubMed and provides a many full text articles as well. PubMed is not specific to pharmacy or medicine, but rather spans a large scope of health sciences, both basic and applied sciences.

There are times when a more focused database is desired. For example, database such as Iowa Drug Information Service (IDIS) or International

Pharmaceutical Abstracts (IPA) are very focused on pharmacy related primary literature. These databases are much smaller than PubMed, but they index some pharmacy specific journals that are not indexed in PubMed, including some very common pharmacy journals such as *Hospital Pharmacy*. IDIS is great in that you get full text access to EVERYTHING in the database, so that's a huge advantage over the other databases. The main downsides to IDIS are that a) it does not interface well with other databases/programs, b) it is not very user-friendly, and c) there is often a time delay between when the article was actually published and when it was indexed in IDIS. IDIS is the smallest of the pharmacy indexing databases, including citations from about 200 different journals.

IPA indexes articles from about 750 healthcare journals, both US journals and English language journals from abroad. IPA is very useful for finding some types of information that you can not find in PubMed, especially information specific only to pharmacy- things like extemporaneous compounding, compatibility, stability, etc.. IPA does interface with a variety of other databases and is available either as a stand-alone product or incorporated into larger packages from various vendors.

EMBASE™ is a European database, similar in many ways to PubMed. EMBASE™ is an excellent database to use in combination with PubMed, as it seems like there is the least amount of overlap between the journals indexed in PubMed and those indexed in EMBASE™. However the main downside of EMBASE™ is cost. It is outrageously expensive and very, very few organizations can justify the high cost for the modest benefit.

How do you manage online journals?

I utilize my favorites list on my browser and utilize electronic tables of content (Etoc). For a MSL, managing data online can get overwhelming depending upon the type of drugs or therapeutic areas the MSLs work with– especially for a primary care focused drug. What might be really beneficial is to set it up a "my NCBI" account in PubMed which will allow you to run a search, save it, then sign up for automatic e-mail updates which provide you with the newly published articles on your desired topic. For example, if you were a MSL for levothyroxine, you could set up your NCBI account in PubMed to search for articles published on levothyroxine then have it automatically re-run your search regularly send you the citations and abstracts for newly published articles on levothyroxine. The MSL could set up different searches with competitors' products for that same disease state.

What news E-tables of content sites do you use (like Firstword[SM] or Healthorbit[TM])?

I do not prefer one to the other. As a drug information generalist, I keep current by reading Journal Watch®. I love Journal Watch®! Instead of looking at all the different E-tables of content, this twice-monthly publication takes the 60 most common medical journals and summarizes the most important articles recently published in those journals. Typically they contain between 20-25 summaries in each newsletter. Journal Watch® also has similar publications for specialty areas like cardiology, oncology, infectious disease, psychiatry, etc. MSLs may find these more focused publications very beneficial. More information can be found online at: www.jwatch.org. The MD subscription rate is $139 for 24 issues, and students, and other allied healthcare professionals can get it for $69 per year.

What about other sites?

I use a lot of the government sites – guidelines.gov, National Library of Medicine (NLM), The Food and Drug Administration (FDA), The Centers for Disease Control (CDC), and The National Institutes of Health (NIH). Some sites are simply not user friendly. You have to learn to use them in order to extract that data you want from them. Another site the NLM has is Toxnet. What I like about this site is that it has a lot of info on reproduction, pregnancy, and lactation with drugs. When talking about safety, this is a good site to review. It is located at: http://Toxnet.nlm.nih.gov/.

What about published versus nonpublished data sources?

Peer-reviewed data is definitely preferred over non-peer reviewed data. Data published or presented through peer-reviewed methods are certainly the most credible. I would suggest starting with published literature like new original research articles (which should be peer-reviewed prior to publishing), as well as the professional organizations (such as the American Heart Association, etc.) for the most reputable information. Other great sources for new information are posters and presentations at national professional meetings. The publication process is a long one and quality information is often presented at national meetings a year or more before the manuscript is prepared, peer-reviewed, and published.

As far as non-published data sources, this arena is more challenging because it is often difficult to tell if the data is factual and unbiased. Additionally, most research is considered proprietary and is not in the public domain until the researchers wish it to be. For various reasons, lots of research is never published. If a new drug is trying to be approved or is newly approved, I will often read the (sometimes huge) transcripts from the FDA advisory committee meetings. In these documents, which can often be found on the FDA webpage, you can find debate/concerns regarding efficacy and safety that are never apparent from reading the package insert or published study data. What were the initial concerns? Were they statistically significant? How did the manufacturer/researchers address the concerns? Rare adverse events aren't always evident from the premarket clinical trials. Also, a MSL needs to know the safety issues with the entire drug class, not just their individual product. The MSL's approach is to be fair-balanced and true to the science. Even if a package insert says one thing, class effects can be an issue for safety and efficacy over time when used to treat larger populations.

Also, the ability to evaluate a study for scientific merit and rigor is important. Study evaluation, interpretation, and application skills are critically important.

For example, it is important to know the difference between relative risk, relative risk reduction, and absolute risk reduction… to know what an odds ratio is, how to interpret a confidence interval, know the number needed to treat (NNT) and what that means… evaluate how drop out rates impact a study, understand how analyzing the data using an intention-to-treat (ITT) analysis is different from analyzing the data from only those subjects that completed the study, etc. These items are critical for the MSL to understand as they review the literature.

MSLs, Metrics and Technology

It is important to utilize medical science liaisons as a company's best continuous source of direct doctor information. When connected with the executive team by good technology, a MSL's field information can play a central role in executive decision making for marketing, launch, and growth strategies. Technology has, for a long time, been in place to support the connection between sales information and home office planning. No focus has been directed toward establishing a proper link between MSLs and the management team. Instead, the systems used by most MSLs today are slightly altered versions of traditional sales/customer relationship management (CRM)-based thinking. Senior management is recognizing more clearly now the untapped value of MSL field intelligence. The obvious next step is to support MSLs with technology matched directly to their work, which can efficiently bring their findings to the executive team.

When it comes to optimal management of MSL efforts, there is a recurring question of qualitative or quantitative metrics. While there is a definite lean away from the quantitative, there is a lot of uncertainty about how to carry out qualitative measurements while retaining sufficient vigor to manage and demonstrate real value.

First, this should not be an either/or question; both kinds of metrics are needed. To avoid the prevalent trap of reach and frequency measures that resemble sales management, it's helpful to ask the question "Why is this measurement being collected?" Is it clearly tied to company strategy for your medical science liaisons? Has it shown close correlation to achieving desired results?

The important steps to take with metrics are 1) track a short list of field activities you think are important, and then 2) test your assumptions by comparing those activities alongside more qualitative measures of results in the field. Where you see good qualitative results, you will be able to validate the direction of your MSL efforts. And where you see less desirable results, you will have an opportunity to refine the direction of your field activities and then run your tests again. Using your measures in this way forms a foundation for management by fact. It provides a good answer to the question "Why should we track our activities?" and creates a cycle of continuous quality improvement for MSL support of your doctors. It is a much more active and useful process than simply setting reach and frequency targets.

About Metrics:
An Interview with Michael Taylor

Michael Taylor is the President and CEO of Akuta Labs. Akuta delivers key opinion leader and MSL strategy, management, and communication technologies to leading biotech, pharmaceutical, and medical device providers. Prior to the creation of Akuta, Michael worked for 10 years in healthcare strategy and technology, including for Trinity Partners, Akuta's sister company. Formed in 1996, Trinity Partners provides state-of-the art approaches and solutions to the health care industry worldwide. Michael graduated with a degree in mind-body medicine from Harvard College.

How have you seen companies quantitatively measure MSL effectiveness?

Quantitative measurement seems to progress in stages. There is a first stage, where very little is tracked. MSL experience might find its way into notebooks, Excel spreadsheets, or even modified customer relationship management (CRM) systems, but this information is generally insufficient for creating good reports or learning from what is happening in the field. In many cases this track-nothing stage can also be quite vulnerable to possible failures in compliance.

The second stage is a swing to the opposite end of the pendulum: track everything. This might lead to some basic demonstration of value, if only through the ability to show a flurry of activity. However, excessive tracking forms an unwelcome administrative burden, and system use (along with any utility) will drop off over time.

The third stage is most interesting. You trust that people are doing their jobs and don't need to monitor every detail, and so move toward tracking the few activities that you suspect are most valuable. At this point, a company can start to test assumptions about how to best direct its MSL force, and refine management to a point where the optimal activities are aimed at a regional or even segmented thought leader level. As field effectiveness improves, a more useful information flow comes in directly from your doctors, supporting your executive planning and decision-making.

How have you seen companies qualitatively measure MSL effectiveness?

Most companies we work with measure MSL efforts against qualitative target outcomes, as well as capture qualitative feedback from field interactions. The challenge is to put in place supporting technology that can wrap qualitative information into easily sharable and analyzable form.

Another key lies not just in what you measure, but also how you measure. Qualitative measures that gauge things like physician satisfaction with your MSL efforts, or awareness levels around your treatment or your competitors' treatments, are great. However, when taken as single snapshots they might generate some interest, but not enough for real learning or well-informed action. A good measurement process should collect data over time, segmented by what you suspect are the most important sources of variation. Then when you go to look at what you've measured, it tells a story that can lead to appropriate response at both the management and individual MSL level.

Qualitative vs. quantitative measurements: which is preferable?

This is not an either/or question. Both are useful on their own. And both can be even more useful when compared alongside each other. The value of activities you are tracking quantitatively should be actively assessed next to the qualitative results that MSLs see in their daily contact with health professionals.

The best measurement programs are ones where 1) you know why you are tracking that metric, and 2) you see specific useful actions that result from that tracking. Nobody should be asked to enter all his or her activities into a black box system and then never see an output. Measurement of MSL field activity should be used in a clear manner that brings better service to their doctors, and to the home office team.

How would you measure the MSL team for optimal performance?

First, it is important to know what are your company's current strategies and key initiatives, and how the MSL team fits within them. How can your MSLs carry out and return with information that is immediately valuable across your operation? When this information is central, you can base your measurement plan on delivering results that will be relevant and reportable to your company's current operations. Yes, some of your measures will remain somewhat generic. But the interesting analyses will come out of comparing those generic activity monitors alongside more specific measures of what is valuable to various interest groups within your company, right now.

Appropriate measurement of MSL efforts also depends on what roles your medical science liaisons are expected to fill, and on where your treatment stands in its lifecycle. Depending on these factors, you might ask questions related to support of clinical trials, satisfaction with responsiveness to IIT requests, development of new key doctor relationships, or shifts in awareness and attitude toward your treatment.

Now with some basic information about current specific aims in hand, pick a small handful of measures. Track your measures through a good MSL-specific system, and connect the right pieces of field data directly back to the MSLs, and to the proper groups within your home office team. The output from measurement is at least as important as the measurement itself, so results should be available in real-time, and in graphic form that is easy to analyze. If it requires lots of work to collect that data, lots of time to get the results, or lots of struggle to decipher a spreadsheet . . . you can have the best measures ever, but they won't end up providing much value to your team.

What systems or platforms currently exist for MSLs?

There are many systems currently used by MSLs, which generally fall into one of two categories: 1) those grounded in a sales/CRM history, or 2) those that are more basic contact management databases. The first category presents difficulties because medical science liaisons aren't sales people . . . so these systems are generally difficult for MSLs to use, don't deliver much that is useful to MSL management, and don't offer much hope for improvement, since MSL interests are commonly the last item on a very long implementation calendar of system change requests.

The second category, contact management databases, offers utility as central platforms for managing KOL efforts, but tends not to deliver good (and definitely not ground-breaking) reporting or analysis of what is happening in the field. These systems can inform different departments of KOL touch-points, but fall short of being able to support executive-level strategic planning or decision-making. This is a substantial shortcoming given MSLs' close continuous contact with doctor and medical information in the field. Medical science liaisons hold a wealth of information about a company's most important customers. A good enterprise system will supply that information in visual, analyzed, and actionable form directly to management.

Akuta Labs created mslConnect® as a full solution for both enterprise-wide thought leader and MSL management. Through our sister company's (Trinity Partners) work with biotech and pharmaceutical companies on commercialization and growth strategies, we were consistently unable to find a single good technology to integrate a proper thought leader database with MSL activity, or to fill the executive team's need for specific information from the field. While we see some KOL software that primarily populates databases through automated search technology, most MSLs are using homegrown solutions (literally not much more than a spreadsheet), or are forced onto CRM solutions that are mainly for sales organizations. We wanted to fix this problem, so we created mslConnect®, which:

1) Provides a complete view of thought leader and MSL strategy, with role or department-specific information interactively delivered across the company,

2) Translates strategy into a straightforward, easy-to-use activity tracking, field-communications, and compliance system,

3) Creates graphics-based real-time reporting of all data elements, allowing for immediate validation and refining of both KOL and MSL strategy.

Managers can drill in and refine factors in their own analyses on the spot; MSLs can see immediately useful feedback on their own doctor interactions. Reports are generally configured to display variation within and among data segments over time; this variation forms a stronger basis for analysis and refinement of strategy than what is offered through traditional snapshots of data.

By giving access to this much information all right in front of you in clear interactive pictures, mslConnect® takes away the head scratching that generally results from traditional CRM, thought leader database systems, spreadsheets, and lengthy slide presentations. You can literally see at every moment exactly what is happening with your doctors, and exactly what needs to be done – both in the field, and in the executive planning room.

Why does the MSL manager need a MSL-specific platform?

You wouldn't expect the same results from a champion cyclist if you put him on a horse...so why would you put an MSL onto a system designed for sales people? Sales teams and science liaisons fill different roles, with different kinds of valuable information to deliver and carry back to their company. The system needs to reflect and support these differences, or the breadth of learning that is available will be lost.

Are there an optimal number of calls for MSLs per month?

Thirty-seven. Just kidding. No, there isn't. How an MSL directs their time depends on things like therapeutic area, company goals, and intentions for their role, as well as segment details like geographic area and doctor type/individual preference. And no matter what the goal, it needs to be tested and refined through comparison to results. This kind of comparative assessment will often lead to small adjustments that can make a big difference. For example, it might matter more how much time you spend with a doctor each quarter than it does how many times you make a call. But if you don't measure and test these assumptions, you'll only be able to guess at how you can improve.

How can MSLs increase their efficiency with metrics?

MSLs should have immediate access to reports of their own activity that compare actions to results. If you can see pictures of what you're doing, how your activities vary over time, and how the results also vary over time, then you have the proper grounding for improvement efforts. This is equally true whether you are talking about individuals or larger groups. Efficiency and effectiveness also improve when the right metrics provide tangible evidence needed to grow your MSL team, and get required budget and resource approvals from executive management.

How does a team optimally report metrics and efficiency?

First it's important to understand your purpose in context of company and therapeutic area goals. Once a measurement plan is developed and reflected in a system that supports MSL efforts, the most important thing is that the measures get used. So often data is collected and goes nowhere. Nearly everyone has been through the design of great new systems, only to gradually get worn out not just by shortcomings in the technology, but also by never seeing any really useful outcome from the system or data collection – whether personally or for the company.

If it just looks like metrics are used to check up on someone, then systems will never be more than a necessary evil, and the information won't be so useful. If metrics can give something back to you and are obviously there to help the team, then measuring and analysis can actually be fun, systems become really useful . . . and your field team gets to grow continuously more effective in giving executives and doctors what they need.

Chapter 7:
Work + Life = Balance?

The MSL role, like any other, often has challenges between the work/life balance issues. One doesn't truly have separate identities; work and life are interwoven tightly. The MSL role is even more interwoven between work and life, because the office of a MSL is typically at home. In this chapter, a MSL father and mother will share perspectives on how they interpret the work/life balance. It will also include some thoughts on work/life balance.

Erin's Wall of Fame

I have a wall in my office I've entitled 'The Wall of Fame.' I include things that make me smile, or make me think, or things I just love. Rather than share with you all the stiff, stoic wisdom I've seen and heard over the years in relation to work/life balance, I'd prefer to focus on my wall for a moment, as I believe a lot of wisdom is stuck to it. It truly translates to any work that one does, whether or not it is work as a MSL, a mom, or any other profession. So, here are the wall bits of wisdom:

1. **Always ask yourself: did today make a difference?** I have this question on a big red piece of paper on the wall. Did you as a MSL have a great conversation that led to better patient care, or new research ideas today? Outcomes are

challenging for the MSL to quantify. However, by asking a simple question such as this, you as a professional will begin to note the impression your work is leaving on others. A good impression is worth its weight in gold, no matter how much someone wishes to quantify or objectify work.

2. **Share your work.** Just as your thought leaders publish and conduct original research, so should the MSL if permitted by your employer. I keep my nametags from meetings up on my wall of fame. Why? To serve as a reminder to share. Make sure you are giving back to your profession by sharing your knowledge. When you help others, you help yourself and your profession. Mentor. Speak. Present data. Share ideas. Participate.

3. **Keep an open mind.** Travel. See the world. Learn. Talk to others in different lines of work and countries. I keep pictures of places I've traveled to in hopes that it will keep my mind open to new ways of thinking or experiences. Science in general is about experimentation, and most, if not all MSLs are into science. The MSL should be naturally curious and inquisitive. I grew up near Chicago, so the Chicago Bulls back in their day were one of my favorite basketball teams. Phil Jackson formerly of the Bulls said, "Always keep an open mind and a compassionate heart". This quote is also on the wall.

4. **Keep your sense of humor.** The wall of course has many crazy Internet jokes friends have sent to me over the years. The smart ones go on the wall. One of my friends (who also is a MSL) sent me a picture of a prisoner digging out of jail from underground, right into a cesspool. The title is "When You Think You Are Having a Bad Day".[11] It always cracks me up, and is especially good to have around after a less than great day at work. The MSL work in my opinion is one of the best jobs in the world for scientists, but even the best jobs and people can have an occasional bad day.

5. **Have fun.** This goes hand in hand with the previous thought, but it is really important to ask yourself if you are having fun with your work. When it's no longer fun, it's time to try something else, period. Fun to me

is thinking about new ideas, so I have a lot of clippings on my wall that talk about new or off the wall concepts. Whatever you call fun, is it integrated into your current work? Ask yourself that question often.

6. **Surround yourself with beauty.** I also have a lot of art on my wall. One of my favorite artists, James Brandess, always sends me his gallery opening cards and each one of them is on my wall of fame. Who is your favorite artist? What makes you happy? Surround yourself with it, because your work will be more productive when you are happy. (Research has shown this to be true.) This is also true of friends and family, which are also on my wall. Do not forget where you came from, as people along the way have made you the great person that you are and that you will become.

7. **Thank others.** I have a few thank you notes on my wall. One of my return notes from a thank you I wrote was written by Sandy Lerner, the co-founder of Cisco Systems, Inc. She spoke at one of my commencements, and to this day I still think her address was written specifically for me at a critical juncture during my own career path. I asked her about starting businesses and she imparted her wisdom to me personally in her note. If you never thank others, how will they know the impact they had on your life? Also, email is nice, but there's something more genuine and heart felt in a hand written note. Take the time to go low tech on your thank you notes if you can, it makes them that much more tangible and sincere. In a recent Accountemps survey, executives were asked about their preferences for thank you notes, and the majority (52%) responded they preferred a hand-written thank you note.[18] Thank a teacher, a mentor, a manager, or anyone that has a positive influence on your life whenever you can. Thank your critics too. They also have made you a better and stronger professional. This also applies to the customers of the MSL.

8. **Be passionate about what you do.** I have a quote on my wall by Jeffrey Katzenberg that says, "If you don't show up for work on Saturday, don't even

bother coming in on Sunday". Even though I personally believe our society is overworked and we all need more time off, those that have passion for their work will never be able to fully disengage. If you are working within your calling, your work will always be there, rolling around in the back of your mind. The good news is that people that are passionate about their work often say their work doesn't feel like work, it is more of a challenge and fun to them. I have the famous Gandhi quote on my wall that says, "be the change you wish to see in the world." This is passion.

9. **Know thyself.** In one of my past jobs, I had the opportunity to have a picture of my aura taken, and the photo is on my wall. I also have a list of my 5 most important values. Every now and then I glance at my values to see if they have changed. It is really important to know yourself, because knowing who you are, what you love to do, and what you don't love to do, is important to your happiness and career. Careers are like gardens: it is important to care for the flowers or work you love to do, and weed out the stuff you do not enjoy doing. Finally, if you do not know yourself and what you want, how will you go out into the world and get it?

10. **Everything is connected.** I loved the movie *What the Bleep Do We Know!?!*™. In it, there is a subway scene of pictures of ice crystals shown taken by Dr. Masaru Emoto. The water crystals were shown or played different phrases or music before they were frozen and photographed. I have a picture of the 4 different crystals that were each played one of four of Vivaldi's Four Seasons and they actually each look different, like the 4 different seasons.[12] Dr. Emoto believes everything is connected in the universe, even the water we drink and the air we breathe, much like Jung's and Bohm's theory of the collective unconscious.[13,14] Therefore, it is important to be kind to others, be fair, be reasonable, and act with integrity with everyone and everything in the universe. Not always easy to do, but something to try and live by.

11. **Take risks.** I have a fortune cookie fortune up on the wall that says, "Nothing dared, nothing gained". I also have a quote from Drew Barrymore that states, "If you don't take risks, you'll have a wasted soul". Make sure you consider all the opportunities out there when working with your co-workers, your thought leaders, or your partners in dialogue. Science is risky. But that's part of the fun for so many scientists, taking the risk to find answers to the questions still unknown is what makes the profession of the MSL fun.

The MSL Father - Tim Hill, PharmD, MS

Dr. Hill is a pharmacist/MSL and has 3 boys under age 10. He lives in the Midwest and he was interviewed for his perspectives on being a MSL, father, and experiences as a pharmacist in the various professional settings he has worked in during his career.

Tell us about your background.

I went to pharmacy school and started working for a large retail pharmacy chain and didn't enjoy it – it was too focused on insurance claims rather than patient care. So I decided to then go to graduate school. I obtained my Master's degree in pharmaceutical science with a major in pharmacology. I had one more year to go towards my PhD, but research budgets were tight and full time faculty positions were hard to come by. I had also recently gotten married so I decided to take some time off to reassess my situation. It was at that time that a friend of mine in a hospital in town called and asked if I knew anyone looking for a hospital pharmacist job. I then applied and a year after being a staff pharmacist, the drug information pharmacist position opened up and they hired me because I had a master's degree. Then some of the local hospitals consolidated, so I decided to depart and went into managed care. I did that for 3 years. During managed care I started a nontraditional PharmD program. I then went into a long-term care (LTC) role to exercise my clinical brain. I then completed my PharmD and then went into the MSL role after my rotations.

Tell us more about your responsibilities in managed care.

I did formulary management, verifying and authorizing restricted drugs, did physician group calls – we had approximately 20 large physician groups and we consulted with them regarding their drug utilization under a dual risk program. We helped them try to manage their pharmacy utilization – of formulary drugs and more generics. We also had special programs. We started to get into Medicare/Medicaid programs with elderly patients and tried to develop the plan for optimal utilization of drugs. I did this work along with a couple of nurse managers.

And from managed care you went into…

Long-term care (LTC). My responsibilities included: patient chart reviews (accuracy and correct reviews for appropriateness on a monthly basis – diagnosis needed to match the drugs). We also made suggestions for monitoring. Also suggested to physicians drug regimen changes in elderly patients (for example, took them off as needed diphenhydramine for sleep.)

And then from Long-term care you then went into the MSL role?

Yes.

Tell us how you found out about the MSL role.

I dealt with a lot of MSLs when I was in hospital in the drug information position as well as the managed care position and thought it was a really cool job. I thought I could do the job as well. A lot of headhunters called me regarding MSL positions, but I didn't have a doctorate at the time. That was a primary motivator for me to go back to school.

What do you like most about being a MSL?

The best thing about the job is that you can pretty much be your own boss. You have to get the job done, but you make the hours and control your schedule.

And that's relevant to being a MSL and father?

Definitely. If I have to pick up my kids from school that day and I'm in town, I go do it. I can also schedule around 6 pm baseball practice.

Do you travel a lot for your work?

I have 2 states, Michigan and Ohio. That's the other great thing about being a MSL; you can fix your travel schedule to meet the needs of your family life. I usually drive most of my territory. What I try to do is schedule my appointments between 9-2 so I leave my house by 6 and I'm home by 6.

What do you like least about the MSL role?

The repetitive nature of it. It's challenging to do the job when there's nothing new and cutting edge – no new data, for example.

What special areas do you think you might move into after the MSL role?

This is it for me! I may consider MSL management, but I have no relocation challenges.

A lot of MSLs struggle with that.

Yes, I wouldn't consider relocating, as I'm very rooted in my local community. If your skill set fits other opportunities, you will usually have to move to the home office. For people who are willing to relocate, this job is an excellent stepping-stone.

Did you already have your children when you became a MSL?

Yes, 2. But I had a smaller territory and the territory size played a role in positions when I was interviewing. There were great opportunities open to me that, had I not had children, I would have pursued more vigorously. For example, one company only needed 2 MSLs to cover the entire country and would promote these positions in a year or two, but I couldn't consider it with a family. So, being a dad influenced the type of job that I was looking for as a MSL in terms of territory.

And you feel you have a good work/life balance currently covering 2 states?

Yes.

And your wife works full time as well, correct?

Yes.

What advice would you give to new MSLs/fathers?

If you want to be involved in your kids' lives, you have to be there. If you have a 6-state territory, it is going to be a challenge to not have overnights. If you are looking to get into that type of job and have a large territory, you probably want to pare that down if you can. That's what makes it possible for me – I live in an area of high density of thought leaders and I'm lucky to have just a 2 state territory.

Did you take time off when you had your kids?

Yes – all 3 times I took vacation and/or paternal leave.

Can you compare and contrast how the LTC and MSL jobs are different and the same?

They are different in that LTC you are dealing directly with patients and patient care. You are making recommendations on a specific patient and interacting with them. As a MSL, you are not directly involved with patients or patient care – it's indirectly involved as you are sharing proper utilization of a drug. The jobs are both similar in that they both allow you the freedom to choose your own hours. You don't have specific times to do chart reviews. A patient load for a LTC pharmacist is 1200-1400 beds. In that month you would need to review all charts. You can review them late at night or early in the morning.

Let's switch gears and talk about mentoring a bit. Did you ever have any good mentors?

I've had people I've looked up to, but never had a strict mentor that took me under their wing and told me how to do my job or how to do my job better. I think it would be helpful. In the long term care role, if you have a good clinical background, and if you don't 'get' the job, you need to look elsewhere. You have to be able to work independently in long term care as a consultant, much like a MSL.

As a former drug information pharmacist at a hospital, could you speak to the benefits of working with the MSLs when you were in that position and what additional value they brought to your position?

The MSLs were tremendously valuable to the drug information pharmacist at a hospital. They can bring new and cutting edge information that other parts or people within the company can't at times. Also, when I would ask a specific question, the MSLs were valuable in bringing to light the studies or case reports that were conducted or published in a particular therapeutic area. I always tried to have the MSL contacts for drugs, particularly new drugs, because they had the information that I would be asked.

Erin Albert and Cathleen Sass

Working as a MSL/Mom: Interview with Carole Carter-Olkowski

After graduating with an Allied Health Degree in Radiologic Technology, Carole Carter-Olkowski worked in the Cardiac Catheterization Lab. While working at the hospital, she obtained a Bachelors degree in Business Management, Marketing. Then Carole began her sales career in pharmaceuticals. After working in sales for five years, she began to work as a Senior Medical Science Liaison. Carole has been in the industry for fifteen years and has worked as a MSL for the past 10 years. She is also married with an 11 year old daughter.

Which came first, mom, or the MSL role?

The mom came first for me. My daughter was an infant when I was interviewing for the MSL job. I was in sales – and now I've been in the MSL role for 10 years. My daughter just turned 11.

What were the challenges of being a MSL and raising a child?

You have to be more focused with your time as a MSL when you have a young child. For me, as a young mother, I constantly had to make sure I was extremely focused during my office time. When I traveled, I had to be organized to have the house and my daughter's care ready to go. The biggest skill to be able to adapt is time management. You have to be precise and try not to duplicate effort. When I was in pharmaceutical sales, there was office time and I found myself doing things on my time – at night, on weekends, etc. It wasn't a big deal then, because you don't really care about how you spend your time. When you have a young child, you are no longer willing to give up the weekend time and evening time to do the work, so efficiency during your work time must be maximized.

What about your territory size and what you would suggest to MSL/Moms in regards to territory?

Well, for one thing, I think that when you have young children and you are interested in the MSL job, in order to ensure success, you need a smaller territory. It is so easy to get overwhelmed in this job and if you are thrown into a huge territory or have multiple territory issues, you are setting yourself up for failure. It is so important for time management to have a smaller territory when you have small children. I've had small, medium and the entire country territories – so I can definitely testify to the fact that young moms need a smaller, more manageable territory. If you have one or two states, that is fine. Travel is still time consuming. Business and national meetings take up time too. Beyond a couple of states, the new MSL mom could be in trouble.

What advice would you have to a MSL that is considering becoming a mom?

Well, I would suggest if you are considering motherhood, first, that you have a workable territory. Also, I would say if you don't already have that mindset to begin your day early, you need to adopt it. It's not really an 8 hour a day job, but you don't want to sign your life away either. Go in with the mindset that you really need to find ways to short cut your day and not add on to it. I'm an early riser, even when my child was young, I would get out of bed really early and front load my day, that way I wouldn't feel overwhelmed and the mental tugging to go back to the computer or the office at night. If you adopt this mindset early on, it will be helpful. Also, if possible, consider childcare at home. There is day care, but a positive thing for me and my child was to have a nanny come to my home. You can eliminate traffic and time spent on getting the child ready to go to day care. I was up early, I knew my nanny was coming, if I had conference calls, I had no rushing concerns. I think it just helps to consider those things when you are having a child. You need to really think about child care early. My husband's job takes him out everyday too, so I knew I would be the one to determine how to alleviate the stress for me.

How far in advance do you need to start looking for daycare?

I started looking for daycare beforehand – before my daughter was born – almost a year in advance, because I already knew (and being in sales then) I wanted my daughter at home and I knew I wanted a MSL job. So, I planned ahead of time to figure out the best way to realize what I desired. One of my colleagues had already become pregnant as a MSL. As soon as she found out she was pregnant, she started looking for a nanny. Even really good daycare can have long waiting lists. You need a whole year to look for proper care. There are some very good websites for finding a nanny. Three websites a friend of mine used are: www.nannies4hire.com, www.perfectmatchnannies.com, and www.enannysource.com.

For me, it was a matter of word of mouth. My neighbor's sister had babysat for my stepdaughters, and she was not employed at the time. She ended up doing her own odd jobs and she babysat for us in the summer. After trying her out, I knew her work, and then asked her if she would sit full time and she agreed. If you don't have the advantage of word of mouth, you need to have some very reputable websites and agencies to help you find the perfect care for you and your child.

What methods of childcare have you used (nanny, daycare) and what worked best for you?

I had a nanny for my daughter. One period, as she became older (when she was 20 months), I decided she needed some part time interaction. I then kept the nanny for 3 days a week, and sent her to daycare for 2 days per week. The difference was staggering – the care she received at home was far superior to the care she received in daycare. I realize daycare is good for interaction, but it was suboptimal for me. I had the nanny come back full time. As an infant, I definitely recommend childcare at home. Also, when I worked at home, it was great to pop in and see her. My office is always away from other parts of the house – but to come up for lunch and see her was awesome. I think by having the child in the home also takes away a bit of the guilt of the working mother, especially for first time mothers. It definitely alleviates the guilt.

At age 3, I had her in preschool for half days. The preschool was in the area we lived in. It was wonderful and I utilized it for half the day and the nanny for the other half day. She adjusted very well to that and it worked out great.

What surprised you most about doing the dual role of mom and MSL?

I was always organized, but what surprised me was the amount of discipline I had – as far as knowing that my daughter was in house but still be disciplined enough to complete work quickly and accurately. I never could turn off the work role before my daughter was born. I would go back to the computer and do work. Once my daughter was born, it became easier to close the office door and not go back in at the end of the day. I sometimes went back in on nights and weekends, but not after my daughter was born, which was surprising.

What do you like best about being a MSL?

The part that I enjoy the best is the interaction with physicians. You aren't detailing or selling, you are talking about disease state, therapeutics, and clinical information. I love that part of the job. The travel and the additional corporate meetings are a challenge. Travel now is unbearable at times. But the interactions make it worth it.

What is your biggest challenge as a working MSL mom?

Obviously, the biggest challenge is the travel while trying to raise a child, and the size of my territory. My territory is huge. My challenge is being able to gain the feeling of being caught up with my work. However, I will never be caught up – there is always something that needs to be done, and I realize that now. Also, being able to have and enjoy my personal commitments is important me. My job does not define me. It provides a means for my family to live with a certain lifestyle. It's important to enjoy one's personal life. I have several other interests. Striking a balance between work and life is a constant challenge!

Working Solo - It's Really A Team Effort

In closing this chapter, it is important to note that no man, nor MSL, is an island. The MSL does not work alone in the field. It is important not only for the MSL to interact with their field based customers, but also it is also important for the MSL to connect with the company through teams and within work groups. Typically, MSLs work with several different internal functions and departments within the pharmaceutical company. The following is a list of internal partners that the MSL can interact and interface with in the company:

Clinical Research and Operations - Understanding the research pipeline is critical for the MSL. The investigator-initiated trial process also passes through clinical operations.

Statistics - A good statistician is worth their weight in gold to a MSL. Many MSLs seek to improve this particular skill set, as understanding the subtleties of clinical trial data analysis is critical to presenting the data effectively. If you as a MSL do not know your internal statistician, they are a great person to get to know. If you are a MSL manager, get to know the statisticians at your company; they may be willing to help train your team.

Medical Information - Some companies call this technical support, medical communication, or scientific information. Whatever your company calls it, it is the department that crafts responses to unsolicited requests for information via medical letters. Knowing the content of all medical letters is a critical skill for the MSL; hence, the MSL should interact on a frequent basis with Medical Information.

Medical Communications - Some companies have medical communications as part of medical information, some separate the function out. However, the medical communication professionals may craft supporting materials and provide educational materials to health care professionals via speaker training or development, or via unsolicited requests. This department, whether separate from medical information or not, is another critical relationship the MSL must form.

Managed Care - Many MSLs provide technical and scientific support to managed care account directors or executives. MSLs can provide scientific formulary presentations upon request from managed care companies.

Sales - Even though this may be a controversial relationship, the MSL must know what is going on with their sales management counterparts in the field in terms of educational and therapeutic initiatives. Also, the MSL can provide technical support and education to the field sales teams when permitted by the company. While in years past, the sales representatives would go on joint calls with the MSL, this is no longer practiced. However, at a minimum, the sales management team and the MSL team should interact to understand needs of the customers.

Marketing - MSLs know the marketplace by talking to thought leaders. They therefore can provide valuable information back into the company. Marketing may need some critical information from the thought leaders that only the MSL can quickly gather.

Executive Management - Although it is not always normal practice, the MSL should interact with executive management. The relationship between the MSL and executive management team is one of the most under utilized relationships and therefore the largest opportunity for MSL teams within the industry. Not only do the MSLs get the opportunity to demonstrate their value to the leaders of the company, but also the leaders of the company have a direct access to the thought leaders within whatever therapeutic area the MSL is working. Managers or directors of MSLs should take each and every opportunity to get the MSL team talking to executive management. Abraham Lincoln believed in the principle of getting out of the office and circulating with the troops.[15] Chief Executive Officers that do not understand the foremost concerns of the leaders in a therapeutic area do not have a firm grasp on their business.

Medical Affairs - At some companies, MSL teams report in to medical affairs. Others do not have this luxury. However, understanding how medical affairs fits into the bigger picture for the MSL is important. (For example, does your medical affairs team report up through research and development, or marketing, or operations?) Whatever setup exists, the MSL will be communicating with the members of this department frequently.

Legal/Regulatory - The legal/regulatory environment exists and is unavoidable for the pharmaceutical industry. The MSL may interact with the legal/regulatory departments on contracting issues for IIRs, for in house clinical trials, or for navigating the environment in which the MSL operates (refer to the chapter on legal/regulatory issues for more details).

Business Development - Rarely, but sometimes, the MSL stumbles upon an in-licensing opportunity for the company. If that becomes the case, the MSL may interact with the business development team.

Public Relations/Media Relations - The MSL may need to provide links between their thought leaders and the corporate home office for media and public relation opportunities. Also, MSLs are not working from the home office and may need to stay abreast of the press releases and information supplied by the company.

Professional Relations - A few companies have departments that only work with allied health care professionals for educational projects, support continuing education endeavors, and perform other non-thought leader related work at a national or international level. If the MSL calls on allied healthcare professionals, it may be important to link them to the person or department at the home office in charge of professional relations.

Training & Development - MSLs can train themselves, each other, sales, or attend courses. Most companies have a department that handles training and development for the sales division at a minimum. The MSL can be involved in developing training materials.

Competitive Intelligence (CI) - The MSL can provide external information on competitors or the industry back to the company. Larger companies often have a formal competitive intelligence division. Sometimes, CI is linked with the business development group at a company.

Research and Development - There should be dual channel of communication between research and development and the MSL. The MSL needs to know about product line extensions, and new products, or novel mechanisms so they can ensure not to duplicate effort with investigator research.

Preclinical Research - This may fall under the umbrella of research and development. Pharmacology basics are also important to the MSL. Many drugs are approved and show clinical efficacy, yet the mechanisms of action are unknown. It is important for the MSL to know the theories or hypotheses on mechanisms of action for the drugs they work with.

Market Research - Again, the MSL hears the voice of the thought leader on a regular basis and this information should be reported back to the company.

Health Economics and Outcomes - HECON or health economic data is becoming as important as clinical research data by many formulary decision makers and managed care plans. Therefore, research needs to be conducted regarding economics in addition to the clinical aspects. A few companies have HECON MSLs.

Pharmacovigilance - Every company is required to have a product safety and pharmacovigilance department. Pharmacovigilance collects all drug safety adverse events information from patients, healthcare providers, and anyone else that reports them to the company. MSLs, as agents of the companies they represent, are not exempt from this. They too are required by law to report any adverse event information they receive from healthcare professionals on their drugs. There are also serious adverse events that need to be reported within a certain timeframe as required by law. If you, as a MSL, have not received training on adverse experience reporting, you might want to suggest that as a topic for your next training meeting.

There are countless other departments based upon the size and scope of the company and therapeutic area the MSL works within. It is good to know that although the MSL works alone in the field, he or she is backed by many professionals at the home office that can support and assist him or her. In turn, the MSL can provide valuable information back to the company, serving as the eyes and ears of the company in the field.

Part III:
Post MSL - What's Next?

Chapter 8:
Job Hopping

As previously addressed in this book, the MSL sometimes feels trapped in a career cul-de-sac after several years of work as a MSL. However, the MSL is never stuck. MSLs have a variety of roles to which they can move with their blend of people and technical skills. Appendix B contains a list of 10 real world career paths taken for current or previous MSLs.

One of the most frequent paths MSLs take is to remain a MSL, but move on to a different company, as the novelty of intellectual challenge of learning a new disease state, therapeutic area, or compound can be provided, while retaining the benefits of being a MSL. Since this is a common occurrence with MSLs, we interviewed a recruiter that works with many MSLs considering the move to another company.

The MSL Moving Onward:
Interview with Bryan Vaughn

Bryan Vaughn is a manager of recruiting with Fidelis Biopharm based in Texas. He has been professionally recruiting for 5 years, and specifically has worked with MSLs for over 2 years. He has spoken at several conferences specific to the MSL recruiting process. He can be contacted at Fidelis Biopharm at 972.770.7922, bryan@fidelisbiopharm, or at his website www.mslstaffing.com.

How did you first learn about the MSL role as a recruiter and how long have you been recruiting for MSL positions?

I first heard of the role when I was recruiting pharmacists and allied health professionals. In my career, I transferred from allied health to pharmaceutical, then transferred to MSL positions. I have been recruiting specifically for MSLs for 2 years.

What factor(s) should a current MSL consider while looking at other MSL opportunities?

MSLs must know what they want. What is their number 1 priority? Is it a quality of life issue - such as size of territory? Are they a manager not wanting as much responsibility, or are they a MSL wanting more responsibility? Are they looking for more money? Are they looking to launch a product, or looking to move to a better company with a stronger pipeline? Do they want to totally change therapeutic areas? How are metrics gathered for a current position versus a new position? These are the tough questions MSLs must ask themselves in order to better understand their own needs in order to better consider other opportunities.

You mentioned metrics. What are the ideal metrics for MSLs?

I don't think there is one best fit for all companies when it comes to metrics. For example, one hiring authority doesn't use metrics in oncology such as reach and frequency of calls. But, the gold standard for knowing whether or not a MSL is doing their job is if at a major medical meeting, such as the American Society of Clinical Oncology (ASCO), the key opinion leader strikes up a conversation with the MSL. The manager knows you are doing your job as a MSL if that occurs. MSL managers shouldn't focus on the number of appointments with nothing to talk about. Rather, the MSL manager should be focusing on the relationship.

What is the average tenure of a current MSL?

Tricky question! It really depends upon the company. In my practice, I have seen an average tenure of 3-5 years at one company. Smaller territories and higher pay keeps MSLs longer. Companies can also keep MSLs longer with strong innovative pipelines. To some MSLs, the pipeline itself is a major motivation for staying with a particular company.

You mention salaries - what is a starting pay scale for an entry level MSL?

Entry level starts between $95K-$115K. Companies and professional backgrounds can alter starting salaries. For example, one company can start an offer for a senior MSL at $115K and another company can pay $130K. There is such gray zone. MDs obviously command more money.

What are the trends in MSLs moving to another company?

I see MSLs move for better science, smaller territories, and/or more money.

What about the MSL that wants to move into management?

I see about 85% of all MSLs get promoted from within their current company. It is very difficult for someone without MSL manager experience to become a manager at another company. That individual must have direct reports in order to become a manager in most cases.

What advice do you have for MSLs that aren't really looking for a new or different MSL job? Should they still talk to recruiters when they call?

The best time to look for a job is when you have a job, not when you're unemployed. As a MSL, you know that the position could change tomorrow.

If a product doesn't launch, you might be looking for a job. A company could buy your company. I just want to help my MSL clients find better quality of life in their work. For example, I contacted a MSL in FL (she was covering FL, GA and AL) and traveled about 65% of the time. I had her CV on file for about a year. I called her and asked her if she could make the same amount of money covering part of FL versus 3 states, would she be interested in looking at the position? She of course said yes. That is why it is critical for the individual MSL to understand what is important to himself or herself while working and continuing to seek other opportunities.

What other trends have you seen in the MSL world lately?

I see a trend of most companies going to smaller territories. But it comes down to the individual. If someone wants a bunch of stock options, they would go to a small company, but would have a bigger geography. At small new pharmaceutical companies, MSLs are compensated better because they are taking a risk and are taking bigger geographies than normal.

What roles can the MSL step into after the MSL role?

The MSLs can transition to many other roles. One MSL went from travel as a MSL to in house medical information as a manager. If the MSL role falls under sales & marketing, they can go into management of sales and marketing. If it's strictly medical affairs, they can move up to managers in medical affairs or can become medical directors. I see a lot of MSLs in that niche - some could go to managed care positions. If they have health economics experience they can go into in house health economics position. Some move into clinical research/operations. It depends upon each person's unique credentials, backgrounds, and interests.

What do you think the future is for MSLs, expansion or cutbacks, based upon your working with companies?

I think the number of MSLs in the field right now is the tip of the iceberg as compared to the number of MSLs in the field 10 years from now. From hiring managers I've spoken with, they are going to be cutting sales forces and adding liaison forces.

Will the old pharmaceutical representative job become the new MSL job?

I don't think so. The MSL can carry more clinical information in their arsenals for educational programs and communication. Why are drug companies spending so much on reps for a 5-minute call when MSLs can get 30 minutes to an hour with a physician? I think there will be different levels of MSLs. What I'm also seeing now are national liaisons. You may have 2-4 national liaisons that are only dealing with national and international thought leaders, but also have a team of 35 MSLs. Some companies also will have clinical trial liaisons (CTLs) - they ONLY deal with products before they are launched - all the research on phases I -- III. The clinical trial liaisons will probably be the Mecca for MSLs. They will need, for example, in cardiology someone that has 15-20 years' experience and will be very niched. They will need to be experts in a specific therapeutic area.

Can MSLs move up the corporate ladder faster by considering other MSL positions?

Without a doubt. Obviously when you can show tenure that is a good thing. But hiring authorities won't draw a red flag on someone's CV if they worked for 3 companies within 5 years anymore. That has changed within the past ten years. It's OK for someone to trade jobs for better science. Having the product that no one else has and the strongest pipeline is the most important thing to many MSLs, because MSLs love the science.

If MSLs are working on a product that is 10 years old or about to come off patent, what will happen to those MSLs? There is no more MSL team. If MSLs are open minded on changing therapeutic areas, their opportunities will increase exponentially. The lifecycle of open positions in your territory shifts over time. It only takes 1-2 companies launching new products to start recruiting people with experience. Those recruits will in turn recruit from other companies all the way down to companies that hire entry-level candidates. For example, cardiology right now is hot. People are launching products and they pull from certain pools of MSLs all the way down to MSLs at entry level. There are trends in recruiting certain therapeutic areas. As MSLs get more open minded, as long as they have MSL experience and would like to consider other therapeutic areas, there are more opportunities for the individual.

Are there any therapeutic areas that people remain within throughout their career?

There are a handful that do this: 1. Oncology - there are people with lifelong experience in oncology. People have been affected by someone with cancer and are passionate about that science. I have never placed an oncology MSL in another therapeutic area. 2. Virology - HIV/AIDS - people don't typically move out of this area. Everything else people will cross over into other therapeutic areas.

Moving around therapeutic areas is a good strategy for MSLs that want to become MSL managers as well. If a candidate wants to go into management and has cardiology, infectious disease, and respiratory experience, this experience set can open more doors for a candidate.

Some companies don't care about therapeutic area expertise, as long as they have MSL experience and are people that can manage people. Other organizations require therapeutic area expertise and MSL management.

What advice do you have for people that want to become MSLs but have never worked in the role?

If you're trying to be a MSL, you need to niche in one therapeutic area. You should go through fellowships and/or residencies.[16] If you are in pharmacy school, do a fellowship with a pharmaceutical company or in a therapeutic area you are passionate about. For example, I have one hiring manager that only wants entry-level people and only people that have done entry-level fellowships.

In conclusion, below is a collective list of the top 30 plus things to have experienced within the first 5 years as an MSL. Some are good, some are bad, but all are from the school of hard knocks, where admission is always open and the education is priceless. This list previously appeared in Pharm's monthly newsletter, *InPharML*.

Over 30 Things To Have Experienced By Your 5th Year as a MSL:

Travel
1. Missed a flight.
2. Had a flight cancelled.
3. Been on a plane, in a train, and in an automobile all in one day.
4. Misplaced your car in the massive academic center hospital parking garage.
5. Arrived to your appointment at 8th *Avenue* when it was actually on 8th *Street,* conveniently on the other side of town.

Education
6. Always kept belief in life long learning.
7. Taken a statistics, health economics/outcomes, or clinical trial design class.
8. Attempted to explain the difference between intent to treat (ITT) vs. last observation carried forward (LOCF).

9. Been asked about any head to head trials between products you work around and the major competitors.
10. Explained to your family exactly what it is that you do, repeatedly.
11. Explained the differences between what a MSL does vs. a drug representative at least 20 times.

Thought Leaders
12. Had a thought leader that refused to meet with you.
13. Had an appointment cancelled last minute due to an emergency--usually after traveling at least 3 hours.
14. Managed damage control between a thought leader and the company (usually for an incident that occurred before your time).
15. Had at least one IIT approved by the company for a thought leader you work with in your region.
16. Delivered, tactfully, more than one IIT not approved by the company to a thought leader in your region.

Shared Expertise
17. Mentored or precepted a student or someone interested in becoming a MSL in your company, and outside your company.
18. Shared expertise in a presentation with the MSL team internally.
19. Presented on MSL work or area of expertise across the industry (presentation, poster, or publication).
20. Led an internal project that provided value added benefit to the MSL team or the company.
21. Contributed time or expertise to at least one professional society (DIA, HBA, ACCP, APhA, etc.) and met other MSLs at other companies.
22. Trained sales reps on clinical data.
23. Developed at least one slide set to be used by the MSL team.

Technology

24. Had at least one company computer meltdown, usually right before a critical presentation.
25. Learned Microsoft PowerPoint® advanced features.
26. Experienced a period of separation anxiety with brand new Blackberry®, Treo™, or PDA.
27. Found an obscure journal article that had to be translated in order to be read.

The Company

28. Went to headquarters for the company and stayed for at least 2 weeks on a special assignment or project.
29. During time at corporate, met and learned from other departments within the company previously unknown to you.
30. Met the CEO of your company.
31. (Respectfully) disagreed with your manager at least once.

Chapter 9: Management of MSLs

Another career option for MSLs is to be promoted to a MSL management position. In this chapter, two different managers of MSLs are interviewed in order for them to share their perspectives of managing MSLs. One is a manager of MSLs in a contract MSL setting, and the other was a manager of MSLs in a small pharmaceutical/biotech as well as large pharmaceutical company.

Medical Science Liaison Outsourcing: Interview with Kyle Kennedy

Kyle P. Kennedy has recently started The Medical Affairs Company, LLC in affiliation with Daniel J. Leonard, M.S. at Scientific Commercialization LLC. The Medical Affairs Company, LLC (TMAC) provides its clients with an impressive compliment of both strategic and tactical medical science liaison program support services. As a Contract Medical Organization (CMO), TMAC specializes in the recruitment, deployment, training and management of MSL teams. Kyle has over eighteen years of experience in the health care industry, the last six dedicated to Medical Affairs with another contract medical organization. In his former role as Executive Vice President, Kyle was primarily responsible for identifying the needs of existing and potential MSL clients, developing solutions, overseeing

MSL programs, vice presidents and directors, and maintaining company-client relationships.

Kyle's MSL management experience encompasses a number of specialty therapeutic areas, spanning activities occurring in the pre-launch, peri-launch and commercialization phases. This experience, coupled with Kyle's intimate knowledge of the MSL position, enables him to identify healthcare professionals possessing the critical skill sets necessary for excelling as a MSL. Kyle has held several management positions in pharmaceutical field sales and Medical Affairs including District Manager, Regional Business Director, and Manager of MSL Programs. He gained his breadth of experience over a 12-year period at a medium sized pharmaceutical company. Kyle has experience in numerous therapeutic areas including pain, cardiovascular, endocrinology, gastroenterology, psychiatry, neurology, rheumatology, urology, respiratory and virology. We talk to him below about MSL contract for hire teams.

How did you get into MSL management?

First off, I started in the pharmaceutical industry in 1989 and took a traditional sales route. I was a sales representative, a hospital representative, and then a district sales manager. In 1994, the company I worked for had a reorganization between commercial, sales, marketing, and medical and we had choice between a severance package or a newly created position. The position I was given was called professional services associate, which at the time was similar to the today's traditional MSL position. I saw this as a tremendous opportunity for me, given my sales background and lack of a doctorate level degree. Big pharmaceutical companies in 1994 had MSL programs, but it was unusual for small to mid-sized pharmaceutical companies to have MSL programs. Within this organization there was a lack of doctorate level candidates, so my department head began with people who had a strong aptitude for science and interpretation of clinical data. I was a heavy utilizer of medical and clinical information as a sales representative, hospital sales representative, and district manager and therefore was chosen to begin my career as a medical science liaison.

I worked as a MSL for approximately 2 years in the Baltimore/Washington DC area and had the pleasure of working with some very prestigious institutions, such as the National Institutes of Health (NIH) and Johns Hopkins. To this day, the MSL opportunity was one of the best positions I've had within the industry. My strengths were in management and I wanted to get back into management, so I was asked to then relocate to Atlanta to the mid sized pharmaceutical company's headquarters in order to expand the group from 10 MSLs in one therapeutic area to 37 MSLs in 3 therapeutic areas over the course of the next 3 years.

What surprised you the most about managing MSLs?

Managing doctorate level professionals is significantly different than managing sales representatives. Specifically, the doctorate level MSLs were much more cerebral and the biggest surprise was that I found they sometimes were technically brilliant, but lacked common sense. This wasn't necessarily negative, but some MSLs were so laser focused on their area of expertise they had not had enough exposure to the commercial side of the industry, and did not have the skills to develop relationships long term and work with key opinion leaders (KOLs). Many of these health care professionals came from clinical practice settings. I think some struggled with their ability to manage people, relationships, and apply common sense because they were over-analyzing their work and their information.

While it is important to hire candidates with technical and scientific acumen, I still continue to hire and look for people that exhibit common sense and the ability to develop and build relationships. Those skills are equal in importance to their academic achievements.

What made you decide to leave the pharmaceutical company and go into professional development of a contract MSL organization?

The pharmaceutical industry in the late nineties had a challenge of measuring the impact that the MSL teams brought to the commercial organization. I was constantly trying to prove the worth of the MSL organization to the business, which is understandable, but the company didn't always accept the soft metrics we were using. They were looking for more objective metrics - return on investment, sales, etc. I actually went back to sales as a regional business director for another 2 years because of the metrics issue. However, I missed the MSL world. My former medical affairs department head had subsequently left the company to start his own contract MSL company, and I decided to join him and help him build that business.

What should managers of potential MSL groups consider if they are starting from scratch on hiring their own full time employee (FTE) MSLs versus hiring a contract MSL team?

Pharmaceutical companies utilize outsourcing in many of their departments. Whether it's research and development outsourced to contract research organizations, or a marketing department outsourcing projects to vendors, or sales adding contract sales representatives, outsourcing is a standard method of doing business in the industry. Usually, a company wants to outsource and be more efficient without losing critical launch milestones. To state it another way, it's much easier to direct budgets towards people and projects rather than receive approval for internal headcount.

What should managers think about? I think what they need to consider is that having their own FTEs provides them (the manager) some job security. Also, if the company is public, internal resources help the stockholders believe there is an infrastructure. But managers should be aware that contract MSL teams are not inferior to FTE MSL teams. A contract medical organization isn't an inferior organization - it is merely an alternative to doing it all yourself.

Also, many managers do not realize that by hiring a contract MSL team, they are also receiving time and attention from a free consultant and expert within the medical affairs arena. We provide our time and expertise to any field-based contract MSL team's hiring management, which is an added bonus to a newer manager.

From the MSLs' perspective, what are the similarities and differences between being a contracted MSL versus a MSL hired as a FTE with a drug company?

I think the similarities lie more in the job description and the day-to-day work. Contracted MSLs do the same type of work as FTE MSLs....the tasks, duties and interactions are the same. Contracted MSLs are measured just like FTE MSLs. I think for the most part, there are perceived differences in compensation. However, compensation is similar between contractors and FTEs, otherwise contractors wouldn't be able to compete for talent. I've seen a large escalation of MSL salaries within the industry lately. There is also a perception by the MSL candidate that a disadvantage of working for a contractor is that your job is constantly in jeopardy because of the contract. I believe this is a misplaced perception. People tend to focus on the length of a contract. The MSL that works as a FTE for a pharma company only has one product, therapeutic area or job to perform. However, if that specific job disappears or the drug goes off patent, the FTE MSL no longer has a position and can be laid off. On the other hand, if a MSL is contracted and the contract they are working on comes to an end, I may have other projects for them to work on either at the same company or at other companies. In theory, the safer or more secure work sometimes lies within being a CMO MSL rather than a FTE. There is another perceived difference that as a contractor, you may not be treated the same. Contractors may not have the same equity position or ownership in the company as a FTE MSL.

What are the pros and cons of hiring contract MSLs?

The pros: you can often do it more quickly and less expensively than hiring your own group. Cons: You don't own the employee asset. You are able to own them if you wish to internalize them. But there is a perceived idea that contract MSLs are less educated and less experienced or less valuable than internal MSLs. However, in my experience, contract MSL organizations are going after the same type of candidate and competing against each other for the same candidates, so I believe this perception is somewhat misplaced.

Could contract MSL positions be a way for new MSLs to enter the industry?

Absolutely, 100% yes! That is the unique proposition. We find our expertise lies in finding candidate types looking for their first break into industry. Many candidates have the clinical experience and education they needed to be a MSL, but they lack one thing, experience within industry. We are the company that can provide them that opportunity for experience, and our employees consistently get hired away from us by major pharmaceutical companies as FTEs, and those companies value them highly. We hire and train them and the industry sustains them. I think this process has created a tremendous network for us over the course of many years - the people that we gave an opportunity to in industry haven't forgotten us. A lot of these people after 5-10 years are getting into higher-level positions in industry and they are comfortable looking at us as an option within their companies. I have been very fortunate and successful in that many of my former employees have helped me to get into new opportunities with other companies.

A lot of people have said to me, "I'd rather work for a real company rather than a contract medical organization". However, you are just as likely to get laid off from a pharmaceutical company as a contract company. But again, contract companies may have more projects, so in theory, a contract MSL organization may have more stable work. The company the MSLs work for as FTEs only have maybe one or two drugs and therefore can't offer as many options as I can potentially. Another often used analogy is that pharmaceutical companies look at us as the triple AAA farm team to bring up new talent into the big leagues of pharmaceuticals. All of the sudden after training and hiring, the MSLs are ready for the big leagues.

If someone with a doctorate wanted to get into the MSL role, where do you recommend they begin their journey?

By accessing any and every piece of info out there from companies like Pharm, LLC, like Jane Chin's company, to all the contract medical affairs groups out there (such as The Medical Affairs Company, Science Oriented Solutions, Ventiv, Scientific Commercialization, and Scientific Advantage). They should contact each and every one of those companies and pick their brains.

Have you seen part time or job share contracted MSL positions?

We have tried both in the past. It takes a Herculean effort to get companies to agree to it, because inevitably they feel they don't get the same delivery or background of information. In terms of job sharing, the company ultimately prefers one MSL to another. It is a difficult thing to do.

Companies want a finite number of people dedicated to the project and have done some type of thought leader mapping and rarely do I see them come up with the remedy of part time MSLs. The issue is that it makes great sense individually but the organizations have a difficult time wrapping their arms around that idea.

You left one contract MSL company to begin your own contract MSL company. What factors did you personally consider to start up your own company?

I was tired of working for the man who worked for the man. A lot of people don't understand the value of MSLs, and it is frustrating to constantly provide justification of their existence to senior management. Constantly showing the return on investment was a challenge. I'm more interested in someone that already believes in the concept and has the freedom to develop a team because they already know that there's a great return on investment for MSL teams. As the industry changes, there is more and more need for fair balanced, accurate information. The customers are spreading to other segments and therefore have different needs. I believe so much in the concept I was tired of selling it up the line to people that weren't really interested in understanding, so I realized that if I could build an organization of my own, I wouldn't have those issues and restraints, and more importantly, have the ability to meet the needs of the customers I serve.

Have you surveyed thought leaders to see whether or not they perceive differences between contracted MSLs and FTE company MSLs?

I have done some surveying in the past. Interestingly, most of the time the thought leaders are not aware at all that the MSLs are contractors; they assume the employee is full time with the company they represent. I have seen a trend of trying to change business cards to say 'in affiliation with' or 'contractor/vendor of' due to perceived legal risks and in the interest of full disclosure, but I don't necessarily agree there are additional legal risks in contract employees having the same card as a full-time employee of the company. I think the concern is the co-employment issues as in the past, such as the Microsoft® contractor case[17]. But in the contracts I write with companies, it is clearly stated the contractors are FTEs of the contracting company, not the pharmaceutical company.

Do you believe the MSL role is here to stay?

The MSL role is here to stay and is growing. I read something late in 2006 on how MSL programs are disappearing, but I contacted the publishers of that article to ask them where they received their data, because I do not agree with this at all. I am in front of numerous companies and speaking at conferences and more and more companies are starting up this role. I think in terms of regulatory and compliance guidelines that there are changes to the MSL role, but the work is needed now more than ever. Pharmaceutical companies have a lot of information to provide in a fair balanced way to their customers. The traditional MSL job (advocacy, training, education, managed market support, investigator-initiated trial development, and attending congresses) is changing. Now what I'm seeing (more so in bigger pharmaceutical companies) are companies starting to specialize the work of the traditional MSL. For example, instead of MSLs performing all 5 functions, some only do investigator-initiated trial work (i.e. clinical trial liaisons). Health outcomes MSLs are quickly growing, as is the field based medical information scientists that only respond to unsolicited requests for information working from home rather than at corporate headquarters. So, I am seeing specialty MSLs as an emerging trend.

Kyle can be contacted at: The Medical Affairs Company
www.themedicalaffairscompany.com
678.881.0872 phone
info@tmacmail.com

Interview with a MSL Manager: Susan E. Malecha, PharmD, MBA

Susan Malecha has over fifteen years of pharmaceutical industry experience in new product development, managed markets, strategic branding message creation, medical education, medical affairs, medical science liaison and management experience. She graduated from Butler University in 1988, completed her Doctor of Pharmacy at University of Illinois at Chicago, and earned her MBA from Keller Graduate School of Management. Most recently, she was Senior Director, Field Based Liaisons at a small biopharmaceutical company focused on developing and commercializing innovative therapies in pulmonology and hepatology. Prior to her current position, Dr. Malecha was the Director of Managed Care Clinical Managers and the Director of Medical Education at a large pharmaceutical company, the Director of Cardiovascular Applied Therapeutics, and held various positions in Medical Information and Drug Safety at other pharmaceutical companies. She has held adjunct faculty positions at University of Illinois at Chicago and Midwestern College of Pharmacy. She is an active presenter/lecturer on Medical Affairs topics for the pharmaceutical industry. She has published papers in Pharmacotherapy, Drug Information Journal, Drug Information Association Forum, and American Journal of Pharmaceutical Education.

How did you get involved in managing MSL teams?

I accepted a position in a medium sized pharmaceutical company as a manager of information development. This position was responsible for providing current awareness information and literature searches and other information resources to the Global Medical Directors, Medical Information, and the field-based medical affairs managers. I was hired to connect people with the same information, to provide consistency between the east and west coast, and sometimes global medical affairs staff. This was the pre-Intranet era. In summary, my job was to unite the communication process. It was a great opportunity, as I learned about the role of the field based liaison, what information was needed, and developed the tools (such as slide decks) they needed. I then moved into a medical marketing position that created customer focused scientific information and education tools for the field group, as well as provided support for pipeline products with the commercial teams, including managed care and group purchasing organizations. After a merger, I switched companies and initiated a regional MSL program for a large pharmaceutical company.

How do you select your MSL teams - what skills, characteristics, and strengths do you look for not only in the individual, but also as well, the team overall?

Initially, I understand the business need and company objective for having a field based medical team. I select a team based upon diverse backgrounds, education, experience and therapeutic expertise for the business need. At the individual level, I evaluate organizational skills, listening skills, problem solving, and communication skills. Will this individual be able to show effective listening skills by allowing healthcare professionals to express their opinions about a particular issue? Can this individual adjust scientific communication based upon level of the audience? Can this person work collaboratively with others?

It is also important that the candidate has an understanding of regulatory drug development processes and has knowledge of pharmaceutical industry. All of these attributes I can ascertain in an interview. It's usually a 50/50 mix of people skills to therapeutic knowledge that I seek. With regard to the selection of the team overall, I hire for diversity and complementary backgrounds. I look for different strengths, so that there is a balance in talent and skills within the team.

In many cases, specific therapeutic background or specialization isn't essential - an MD, PharmD or PhD can learn the therapeutic area. There are stellar field based liaisons that have superior communication skills and people skills, perhaps who are allied healthcare professionals that also know the science, yet may not have the 'D' behind their name.

What skills and talents do the best MSLs share?

There are a few fundamentals to being a successful field based liaison. As I mentioned previously, communication, the ability to translate information at different levels, and then develop relationships with all "customers," whether it is the community practicing physician, the study coordinator for a research site, a pharmacist, or a national thought leader, is key. The liaison must have the ability to provide scientific exchange appropriately. Secondly, the field-based liaison has to show expertise in their therapeutic area, knowledge of their products, patient treatment trends, clinical studies, and scientific activities in that area. This lends to credibility. Thirdly, a field based liaison needs to understand the research, drug development and approval process, and be therefore able to contribute to clinical support functions. Fourth, the individual must possess integrity and creative spirit needed to thrive in a changing environment. Fifth, a liaison needs to be emotionally mature. A field based liaison works independently in a face-to-face setting, not at a sheltered corporate office. Emotional intelligence, flexibility and the ability to use good judgment at all times are very important.

MSLs tend to be independent and self-sufficient - how do you get the individual MSLs together to form a cohesive team and share best practices?

I set the tone and expectations for the team right at the beginning. I establish the tone around mutual respect. Everyone is listened to and valued. They are a team and have a common goal they are working toward for the company. Communication is, once again, vital to achieving goals. Through team meetings, conference calls, and informal routes like email and voicemail, field liaisons need to be communicating with each other and the home office. I have paired up a new employee with a more seasoned one or created the buddy system for the first 3 months out in the field and found this really helpful for newer field based liaisons to understand and manage their role and activities.

As a manager of MSLs, how do you keep them interested and motivated in their work? What techniques have worked best? Particularly after the product lifecycle is beyond one-year post launch?

When you first start managing, you have to link the learning of the field liaisons to corporate goals and strategy in their jobs. For example, this is what the company needs to have done (XYZ) and this is what the person has to do in their job locally (XYZ in territory for specific MSL). That means a field liaison may have to customize and be creative to meet the goals of the corporation in a way relevant to their geographic assignment or territory - through appointments with thought leaders, principal investigators, or other professionals. Each field liaison has to know their territory (understand the important accounts, the significant institutions and academic centers, the influential regional thought leaders) and then meet the company objectives within their territories. The most motivating factor is team achievement of goals. There is a personal satisfaction, as well as acknowledgements from the company and others.

For older products, it may be a bit more of a challenge to keep field liaisons stimulated compared to a product's support at Phase III to launch. It is important to follow their specific objectives, yet this may be the time to expand their expertise and take on different endeavors - like a negotiating skills course or a pharmacoeconomics course - something to broaden their knowledge. The liaison is still doing their job, but we are challenging them to think more broadly for the future and stimulate them intellectually.

You have managed different teams on different products in different therapeutic areas - what are the differences and the similarities?

Different therapeutic areas may require a specific expertise or knowledge base, yet the management style is the same. Oncology, for example, is a specialization. People have to be well versed in oncology and field based liaisons rarely move to different therapeutic areas from oncology. Field based clinical positions that support pharmacoeconomics or managed care are trickier as well - different positions and different backgrounds are required. Returning back to the business objectives and goals and managing to that will place the right mix of talent for the company need.

Metrics have been one of the biggest challenges for MSL managers - what works for you?

I manage by objectives, using qualitative and quantitative goals. These have to be company specific. I am not supportive of the frequency model (for example, seeing a thought leader X number of times during a year just to see them). The objective is scientific exchange to the health care professional so he/she can make the best and appropriate treatment decisions for their patients. Visiting that decision maker 10 times in 3 months may not have the impact of a single visit. In one company, the field medical group had four major objectives: 1. To support clinical development, 2. To develop advocates, 3. To provide and support medical education and disease awareness and 4. To provide internal marketing support. One can develop quantitative and qualitative goals around those objectives.

If a new MSL manager were just starting out, what advice would you have for him or her?

Be open to all possibilities and be patient. It is important to listen to your team and understand the expectations of the company on you as a manager. It is also reasonable for new manager to feel challenged in putting all the pieces together – commercial support, clinical support, medical affairs integration, sales training support, field sales, national accounts, company policies, different business units, etc. It eventually all falls together. For a team's value to be seen internally, it may take up to four years to fully integrate the field based team into the corporate model. For an individual MSL, it takes at least 6 months for them to be comfortable in understanding the role and their geography.

Relationships take time to develop too - it is not a sprint race but a longer run. As a team leader, honesty, integrity, and open communication must be present between the field liaison and their manager. There are more roles than just relationships within the field based team. The team leader has to work with internal and external customers as well. The field team leader has to be an advocate, confidant, mediator, organizer, consultant, colleague, arbitrator, and, as team leader, be the protector of the team. There are multiple responsibilities, and all of them are very important.

Chapter 10: The Afterlife

The MSL job is a great one. However, if you have been a MSL for some time and feel trapped in the career cul-de-sac, here are some other suggestions of jobs that might be an interesting next step for you. In many cases, you will need to determine whether or not you are ready to make financial and/or relocation sacrifices in order to make a change. Remember, there is more to life than money; your happiness in doing what you do for most of your waking hours should make you at least semi-happy. Life is far too short to be miserable.

Other career alternatives for MSLs:

Business Development - Nearly every company in pharmaceuticals and biotech has a business development person or group. If your company does, get to know some of these people. They meet new people all the time, assess concepts and compounds and the business validity of a project. Sound familiar to the MSL role? It should, many of the aspects are similar.

Entrepreneur – That annoying little idea in the back of your head that you've been thinking about for a long time …well, what are you waiting for? Go for it! You can find a plethora of small business resources online and with your customers, all you have to do is ask. (See the interview with Dr. Jane Chin at the end of this chapter for her insights in the transition from MSL to entrepreneur.)

Publication Planning – If picking apart clinical trial data and/or writing is what gets your juices flowing, you might want to begin conversations with your company's publication planning group. Many medical writers also freelance from home.

IIT/IIR/Clinical Project Management – Love developing and championing investigator initiated trial concepts? If so, do you have someone on the inside that funnels all these ideas into home office? Get to know him/her. If not, why couldn't this job be created?

Managed Care – There are two sides to the managed care coin. You could, as a former MSL, work for a managed care company as a formulary manager (typically PharmDs or MDs are in this role) or you could work for managed care within your pharmaceutical company. Either way, it is yet another option to help you grow professionally, as you can learn more about how managed care companies make drug formulary decisions.

Medical Advertising Agencies - If you like articulating the features and benefits of a drug, the advertising agency route may be another to consider after the MSL role.

Expert Witness - If you have a PhD, PharmD, or a specific area of expertise, freelancing as an expert witness can be a nice part time or full time career path. Talk to your attorney friends and find out what local outages exist for experts in pharmacology, toxicology, and your therapeutic area(s) of expertise.

Medical Content for Information Technology - If you enjoy medical writing either for the lay public or for other medical or healthcare experts, you could write content for an online medical information portal. Many state and local pharmacist associations also have a need for newsletter construction and content.

Competitive Intelligence (CI) – In some companies, this is a separate department. In others, CI is in business development. Either way, you can get paid to learn about what other companies are up to. What is the cutting edge science in particular therapeutic areas?

Teaching/Academia – The pay isn't the best in the world, but there are perks to academia you can rarely obtain in industry: sabbaticals, down time in the summer, less travel, and the ability to practice clinically part time if you so desire. (See the interview in this chapter with Dr. Scott Stolte who made this transition.)

Health Economics (HECON) – Some companies already have health economic medical liaisons. If your company does, this may be an interesting option for you. Talk to people in health economics or take a class. If your company doesn't have a HECON group, start talking about the benefits to having one and champion the need for outcomes based research in your company.

National Allied Healthcare Liaison – The larger companies have a person fully dedicated to working with professional groups, (pharmacy, nursing, osteopathy, respiratory therapists, etc.). This job is similar to a MSL but the customer is slightly different and sometimes on a more of a national level than regional. Ask about this as an option if you enjoy the interaction with customers on the association side of the business.

Not For Profits – If you have an itch to create and/or lead while staying in the health sciences, working for a not-for-profit may be for you. Talk to some of the presidents of not for profits that you currently call upon to assess their own job satisfaction, the benefits, and the challenges.

Government - Many types of medical professionals are needed in places like the National Institutes for Health (NIH), the Drug Enforcement Agency (DEA), the Food and Drug Administration (FDA), the Division of Drug Marketing, Advertising, and Communications (DDMAC), and other health advocacy/political avenues. If you have a desire to try making a difference at the government level, looking into federal state or local options within the government sector would be an interesting route to explore. If you would like to work with the government but remain on the pharmaceutical company side, another idea would be working for regulatory affairs.

Scientific Sales Training - MSLs moving on in this role can work with sales and marketing to ensure sound medical science is being shared and training is accurate for pharmaceutical sales professionals.

Medical Marketing - MSLs can move into affiliate or global marketing with a pharmaceutical brand and work between the brand and medical, for example, to ensure accuracy and consistency among the science as well as the brand.

There are countless other opportunities for MSLs. Three viewpoints from professionals that moved on from the MSL are provided - Dr. Stolte moved from the MSL role into pharmacy academia, Dr. Chin moved from the MSL role into the role of entrepreneur, and Matt Lewis moved from the MSL role to medical education.

Academic Pharmacy After the MSL role: Interview with Scott Stolte, PharmD

What is your background?

I went to pharmacy school and graduated from Purdue in 1997 with my PharmD. After school, I did a residency in community practice with services and traditional pharmacy community practice (disease management clinics) and then joined the faculty at Shenandoah in the college of pharmacy. I left academia for a while to be a MSL at a pharmaceutical company that worked with diabetes and 2 years later, went back into academic pharmacy.

Where did you hear about the MSL role?

I had a number of colleagues from school that did fellowships and they became involved with pharmaceutical manufacturers. It sounded they did a lot of similar things to what I was doing in academia - teaching, educating, and providing information. I thought it would be great to educate people on the front lines of healthcare. In many ways the roles are very similar.

Did you interview with many companies for MSL roles?

Yes - I actually interviewed with 3 or 4 companies. I considered not only disease states and drugs I would be working with, but I also looked carefully at their mission statements to understand their level of commitment to the disease state that I would be focused upon.

What were your primary responsibilities as a MSL?

Most was providing information about medication (about 80%). The job has since changed. We did do some site identification for clinical trials. The drug I worked with at the time was new and I felt I played a valuable role in providing the initial information regarding the medical information to healthcare providers. I was in this position for nearly 2 years and was satisfied in the position but I decided my passion was in academic pharmacy.

How did you slowly discover your passion for academic pharmacy?

Honestly, I missed the students. There is a certain excitement and feel you get interacting with students on a daily basis that I just didn't have with healthcare professionals. I could see how some might be more engaged when working with the healthcare professionals versus students, but my heart was with the students. There is a feeling where you see the light bulb go on for the first time in the students when they truly understand that it is truly rewarding and knew what I wanted to do.

What about mentoring as a MSL, how does mentoring differ between the academic setting versus the corporate world?

In academia, junior faculty are assigned to senior faculty for a mentoring relationship. In school it is teaching, service, and scholarship. The scholarship component was the most important reason to have a mentor in academia. Mentoring in the MSL position was more challenging, because all of the MSLs started at the same time and we had one manager with 11 MSLs. I think mentoring is critical to people in discovering their best career paths. It is important to have a small ratio of 1:1 or 1:2 for mentoring for it to be optimal.

Was the decision to go from MSL back into academia an overnight decision or always at the back of your mind?

It's a good question, and it was a little bit of both. I missed the university and the interaction with students. However, I found that the MSL position did fulfill a lot of those things I was looking for. The same university I worked for previously (Shenandoah) called me one day and said they missed me too so I reconsidered and decided to go back to academia.

What are the biggest challenges for you in academia?

People in academia are extremely thoughtful. Decisions are made only with significant deliberation. For someone like me who makes decisions on the fly and makes decisions based upon instinct and quick reads, it can be somewhat frustrating. This is something a person should truly consider when thinking about academia. The other thing about academia is that everyone is extremely knowledgeable. Getting through strong opinions can be tricky at times.

If someone were interested in exploring academia from the MSL role, how could they explore this option?

I think the first step they should take is to check in with the local college of pharmacy to start precepting for students' rotations (with the company's permission, of course). There are some people that glamorize teaching, but it is a lot of work. Students need to have their questions answered. The best course of action is starting out slowly and exploring the option.

MSLs struggle on where to go from their positions. What are the career and development opportunities in academic pharmacy?

The career ladders are fairly clear in academia, which is good. There are also a number of opportunities. Most people start in regular non-tenure or tenure track positions. But over time, pharmacists in academia tend to go more toward the service side, or the scholarship side or the teaching side. There are a number of administrative positions in each of those areas that pharmacists can pursue. The great news today is that clinician pharmacists can advance to deans. It can be both at traditional research institutions as well as primarily teaching institutions. One of the other benefits in academia is that professionals can pursue higher-level learning -- by pursing a master's degree or a doctorate.

What are the differences in compensation?

There is no doubt about it - there is a major difference in compensation between academic pharmacy and the MSL role. The median for an academic pharmacist is about $76K per year (2005) - much different than for a MSL. A person really has to consider this when going into an academic role. It's not always about the money, but it is a little bit. One has to be truly passionate and committed to education in order to consider this. Retirement benefits at universities are strong and attractive. The academic pharmacist however can create his or her own schedule and it is very flexible. The MSL role also offers some of that, but the travel isn't as extensive for academic pharmacists. You can control your own travel with meetings; it is all your choice. Most schools also have benefits on lower cost advanced education. Some schools leave the coursework wide open, but others limit it only to one's current profession.

What other requirements are there for academic pharmacists?

Residencies are required by many schools. Some schools will allow for 5-7 years of practice experience in lieu of a residency. There should be some type of postgraduate training. Fellowships are not yet required. Fellowships tend to focus more on research, whereas residencies focus on clinical practice and teaching. Some schools may require board certification.

Can you talk more about board certification?

Many schools of pharmacy are requiring board certification for pharmacy practice faculty. The pharmacotherapy exam is a broad based test to support and demonstrate a broad based knowledge. This may also serve the MSL well to discuss patients with multiple disease states and have clinical conversations with other healthcare practitioners.

What organizations could be valuable to the budding pharmacy academician?

The American Association of Colleges of Pharmacy (AACP) has an interim and annual meeting that is focused on pharmacy education. Some of the universities have a general course on what it is like to be a faculty member. It is very important for anyone considering this to talk one on one with current faculty members to see the advantages and disadvantages.

Entrepreneurialism Post MSL: Interview with Jane Chin, PhD

What is your professional background?

I started working in the biopharmaceutical industry in the early 1990's. I worked in different functions, including analytical research and development (R&D), sales, and medical affairs as a MSL. From there I became an entrepreneur. Medical Science Liaison Institute is one of my companies and is focused on MSL services.

How did you first learn about the MSL role?

I learned about the MSL role when I was a primary care pharma sales representative. I often worked with the hospital representative, and one day, she brought in a MSL. Most of my knowledge of the MSL role is by observing the interactions that the hospital representative had with the MSL. At the time, there was literally no information on the Internet about MSLs that I could research.

Please provide a brief synopsis of your career as a MSL.

A few months into my MSL career, our team was laid off because the FDA had issues with the (now defunct) company's manufacturing facilities. The reality of the "security" of working as a MSL became apparent to me early on in my MSL career. It was a good lesson, but quite a shock when we got the news. This was my first (and last) time being laid off. From there, I worked at a privately held pharmaceutical company. I enjoyed my job even though I had ten states in my geography. Unfortunately after September 11, 2001, travel became a serious work-life issue for me. I transitioned to another company for a smaller geography.

What did you love most about working as a MSL?

When I've asked MSLs this question, many cite intellectual stimulation from interacting with thought leaders as their top motivator in this job. I loved the interactions with my MSL colleagues the most. I remember agonizing over my decision to leave a former position because I liked my MSL colleagues very much. We supported each other, were generous with our time, and we genuinely wanted each other to succeed. I enjoyed interacting with thought leaders as well, although over time, I found myself drawn to learn about the issues that my MSL colleagues were facing.

When did you start thinking about career development beyond the MSL role?

I was thinking about career development before finishing graduate school. I created a career manual for myself when I was a graduate student, so I had been proactive in my career development even before becoming a MSL. I really wasn't sure what I wanted to do beyond the MSL role. I knew I didn't want to manage people, but that seemed to be "the way to go" on the MSL career path. At one point I seriously considered getting a law degree or an MBA. I had no desire to become a lawyer or concrete plans for the MBA, but I looked at these options as a way to challenge myself.

What factors made you decide to become an entrepreneur?

This ties into the previous question about career development. I couldn't decide what I wanted to do for the rest of my life. I wasn't even sure that I wanted to do only one thing for the rest of my life. I wanted freedom to explore different opportunities, including those outside the life science industry. As an entrepreneur, I can design my future as I see fit. Of course, this makes answering the question, "What do you do?" incredibly difficult for me today.

Looking back, what skills did you obtain or sharpen during your tenure as a MSL that are helping you as an entrepreneur today?

I had a lot of practice "traveling", which could be helpful if your ventures require you to be a road warrior. My experience with changing deadlines or long-term projects that are the norm of the MSL profession was useful to an entrepreneurial transition. Since my work through Medical Science Liaison

Institute focuses on MSL programs and issues, my MSL experience helps give me a level of insight that I may not otherwise have.

What advice do you have to give an MSL that is thinking of moving on to being an entrepreneur?

Invest in your personal and professional development, and continue investing in your development. Get used to not knowing where your next mortgage payment (or rent) comes from; this can be tough after earning a six-figure salary for many years. Know why you want to do the next thing you want to do, and if you're not sure, work with someone who can help you figure this out. Know what you want to make of your life before you worry about how you are going to make a living. Make time for fun, and if you can't seem to find time for fun, schedule it.

Jane can be contacted at: Medical Science Liaison Institute
info@msliq.com
(310) 542-5642

Ten Transferable Skills From MSL to Entrepreneur:

1. Ability to solve problems.
2. Ability to interact and alter conversations based upon audience.
3. Ability to network, explore the new and the different, and become a connector of disparate ideas or people.
4. Ability to work independently.
5. Ability to take risks and try new ideas.
6. Speaking skills.
7. Ability to maximize follow up and follow through.
8. Ability to think on your feet.
9. Ability to take the complex and make it simple.
10. Self motivation.

Erin Albert and Cathleen Sass

Life After the MSL Role in Medical Education: Interview with Matt Lewis, MPA

Matt Lewis is currently an Ed.D. student in Interdisciplinary Studies, Columbia University-Teacher's College. His academic interests include adult learning, public health education, and clinical informatics. Matt graduated from Cornell University, cum laude, with a Bachelor's of Science in Molecular and Cellular Biology, and earned his Master's in Public Administration, with distinction, in Health Services Research, at the NYU/Wagner School of Public Service.

Matt has been with his current private pharmaceutical company for over a year, acting as Manager, Continuing Medical Education, where he has responsibility for continuous professional development activities in the Central Nervous System community. Prior to joining his current company, he was with a large consumer products/pharmaceutical company for over seven years. While there, he held a number of positions including those in market research, pharmaceutical sales and scientific affairs. He served as both a medical science liaison and a regional scientific advisory board manager.

How did you get started in industry, and how did you discover the MSL role?

I started in industry after exploring a lot of other alternatives that could utilize the juxtaposition of my educational degrees for real world application. I did not want to be a clinician because I felt it might be too limiting. I felt I could make more of a difference by working in a larger organization with a broader scope. I explored a lot of directions and a couple of key mentors at Cornell, who had life science industry experience, helped me understand more about the opportunities available to graduates. They assisted me in identifying multiple internships in the field of hospital administration, laboratory consulting, license and acquisition, information technology, and market research. This was mainly helpful in ruling out work I did not want to pursue. Licensing and Acquisition (L&A) and intellectual property issues were most intriguing, but I wasn't interested in going to law school. From a practical perspective, the options that remained left me with either the insurance or biopharmaceutical industry. To me, industry felt like the right place to solve a variety of problems and contribute to the public's health while exploring and growing my career.

The first time I learned about the MSL role was when I was an intern at a pharmaceutical/consumer products company in 1999. The MSL role was just being developed around then at that particular company. Someone from the group came and presented to us about what the function looked like -- scientific exchange, relation of lab science to the external world, and what's actually taking place among the experts. Based upon the role as it related to biology and business, it sounded like a perfect role for me. Before becoming a MSL, I paid my dues and learned how to generate revenue and have respect for the sales role at the company (it was my first job there). After finding about the role as an intern, being in a current sales representative role, I decided to strategically plan everything beyond the normal scope of the sales job to build a skill set for the MSL role. I developed my training ability by training reps on the science, terminology, and how to scientifically sell. I also had the opportunity to develop a publication for research and publication review for internal use in order to educate representatives. Before I even left my sales role, I was functioning more as a MSL rather than straight selling. I saw opportunities to bring the science to the sales role, and maximized that skill.

How long were you a MSL?

The MSL role was for 2 and a half years. As I developed into my role, there were challenges from a company perspective on where the MSL could expand from in terms of career development. During my MSL and board manager tenures, I had the opportunity to work with others in CME and publications. But while working in concert with these roles, never did I say to myself, "Oh that's the job I want to do". When I was a MSL, I viewed the board manager role as purely relationship based. At the time, to me, there was no real scientific component of the work itself. That might have been personal bias based upon how I viewed others in the role. It was not how I conceived of the role, and when I was offered the opportunity to step into the board manager role, I decided to re-create it in my own image. The director in the department, whom I had not spoken to until then, approached me to become the board manager in my region. My manager encouraged me to consider the board manager position. The interview was largely a formality at that time. After becoming a board manager, I found I enjoyed the role because I had the ability to pursue scientific projects in a way the MSL role did not allow.

What jobs did you take after the MSL role, and how did the MSL role prepare you to do those assignments more effectively?

I think the most interesting thing about being a MSL was the satisfaction at the end of the day of influencing the influencers. To me, the ability to make a difference and help others to become educated was most interesting. When my thought leaders would talk to fellows or others and have effects positively on patient care by providing new information was what was rewarding for me.

You did not have a doctorate when you worked as a MSL. Do you think the doctorate should be required of MSLs as a baseline for entry into the role?

I do not think a doctorate is required as a baseline entry point for MSLs, or even after. After they have proven to be a good MSL, a doctorate isn't necessary. I disagree with it both in principle and in practice - it is inane.

What skill set do you think is necessary?

The doctorate education appeals to a lot of the basic educational background needs of a MSL, and I see why people would gravitate towards this baseline education. However, some people have the skills to be a great MSL and some don't, regardless of the professional degrees they achieve. I think the essential skills for a MSL are the following: 1. Strategic planning - a lot of PhDs completely lack knowledge on how businesses operate. To have the ability to plan and implement in a business context is a different way of thinking. How the little things add into the overall big picture is important. 2. Effective communication - the MSL must excel at all forms of communication, both oral and written. An individual must have the ability to professionally articulate ideas and concepts in order to be received and be taken professionally. 3. Technical mastery - some people are great with understanding technical data, but do a lousy job of explaining the relevance and appropriateness of the data in the larger context to the audience to which they are presenting. For example, a MSL must understand the data around a product, but they also must have an understanding of information technology and how the US healthcare system is evolving. Items such as health policy and quality improvement are important in relation to a particular compound, and most importantly, what the big picture looks like and how whatever data the MSL is presenting fits into the big schema. To become an expert generalist is important for the function. 4. Project management - it is important to have timelines attached to milestones and walk a strategic and technical path. There has to be a plan for the MSL's work.

What was your favorite thing about being a MSL?

The thing that drove me was influence. It wasn't the salary or the flexibility or professional status. Those elements are all nice, but what interested me was the ability to make a direct influence with a thought leader, who in turn provided information or education to evolve healthcare and ultimately, provide better patient care.

What was the most frustrating thing about being a MSL?

I was a MSL technically twice. The first time, the largest frustration for me was the lack of understanding and appreciation for the MSL role internally by other departments.

The major internal groups lacked understanding and recognition of the value that MSLs could provide and serve information not only to external thought leaders, but also work with other internal departments. The second time I was a MSL, the biggest challenge was the feeling of taking a step backwards and not having as much educational influence as I had being a board manager. There was less of a feeling of strategic impact than I had previously had as a board manager.

Some view the MSL role as a 'career cul-de-sac' or a dead end job. Do you agree, or disagree, and why?

I strongly disagree. I don't think any role is a career cul-de-sac. Every job is what the person makes of it. I've had 7 different jobs in as many years - because I constantly needed a new challenge and saw connections between current and future responsibilities in ways that were not outlined for me. You have to take the initiative and recognize what you enjoy doing vs. what you don't enjoy doing. Sometimes I think some within the MSL role expect the academic track to exist within the business world and the company therefore owes the scientific professional career paths. Because the fact they are from clinical or academics and now in the business sector, they sometimes still expect the academic track to exist within the MSL role. In academia, there appears to be a distinct career track - tenure - to promote people. However, in the business world, one has to be cognizant that the clear career development path does not always exist. One has to manage his or her own career, manage their manager, so to speak. Does your manager know what you want to do as a next career step? Also, one has to be able to navigate the lateral and vertical options for career development. A MSL or employee needs to proactively seek mentors. This takes work, and many people either give up or do not do anything to progress their careers.

What advice would you give someone that is currently a MSL that might be considering other career alternatives?

I would say that there are 2 things that matter for any job: 1. Network, network, network. Treat all interactions like potential job interviews. 2. Explore what other functions exist and what is compelling within them, and then maximize those skills needed for the next job within the current MSL role. For example, if a MSL loves to train others, do more training. Work with the internal training department, pursue educational companies and adult learning professionals and talk to them. There are so many opportunities for other functions in medical and marketing. The individual MSL just has to be proactive in seeking out his or her own career path. One should self assess, know strengths and weaknesses, and find mentors.

Do you think the MSL position will endure?

I think the function will certainly last. There will always be some intermediary between scientific affairs (clinical operations and publications) and a thought leader. I don't know in the future if the function will live within medical affairs or the people doing the job will be called MSLs. Some other groups within a company do a better job of providing objective tangible value via metrics. The day-to-day job of the MSL is to share data - but the value the MSL provides back to the organization is the key. The MSL will endure in the organizations that can figure out the proper metrics for this role and clearly demonstrate the value. If they cannot, the role may take on another name or be in another functional area of the company, simply due to the fact that the other area figured out how to more effectively present and demonstrate the value of the function back to the organization.

You left the MSL role and moved into education. What are you doing now?

I am a manager of continuing medical education (CME) at a pharmaceutical company. Half of my current colleagues were MSLs at one point in their careers as well. So, many MSLs go from the MSL world into the CME world. There are many people in medical education with a MSL skill set. The biggest additional skill I needed to work on in my new role is enhancing the ability to manage projects and stop trying to get into the details of the data. The MSL role is to share data to influence the influencers. Now, I need to step away from this temptation and focus on the bigger picture of educational development and project management.

Chapter 11:
The Future of the MSL Role

Where is the MSL headed? A previous chapter covered the concept of specialization of the MSL. This chapter contains an interview with Brian Best on an evolving role for MSLs: the clinical trial liaison. Finally, Dr. Stan Bernard, interviewed previously on the history of the MSL, provides his conclusions regarding the future of MSLs.

The Clinical Trial Liaison:
Interview with Brian Best

Brian Best has worked within the pharmaceutical industry for over 15 years and has held various leadership positions within marketing, sales, business development, clinical affairs, and medical affairs. In one of his previous roles, Brian oversaw a national clinical trial liaison team focused on the successful enrollment of a pivotal Phase III study. In this interview, we focus on Brian's expertise and strong recommendation for companies to consider a different breed of employee within the pharmaceutical industry: the Clinical Trial Liaison (CTL). The CTL role blends some responsibilities from MSLs on the commercial side as well as responsibilities from clinical research associates (CRAs) on the preclinical and clinical portion of the business to aid in and streamline clinical drug development programs.

Erin Albert and Cathleen Sass

How did you become interested in the utilization of some of the MSL functions in clinical operations/research?

While serving as the MSL Director at a small pharmaceutical company, I leveraged the MSL group to reengage more than 50 investigator-initiated trials (IITs) that had been "lost to follow-up" during the merger between the company I was employed by and another company. It was during this period of IIT revitalization that I discovered that MSLs could have a dramatic effect on principal investigator (PI) motivation and study coordinator (SC) activities. I took what I learned of this aspect of MSL responsibility in research and transferred it to a new opportunity at a different start-up company, which was having a difficult time enrolling a phase III study in cardiogenic shock. I built a CTL group from the ground up to be effective at one thing above and beyond all else: successful enrollment of a pivotal, phase 3 study. The results were impressive, and I have continued to utilize this function in clinical research. We are building a new CTL organization at my current company, where a MSL group already exists.

What is a CTL, and how are their roles different from that of a MSL?

MSLs are focused on scientific support for commercialized products and CTLs are focused on clinical development of pre-commercial drugs. The opportunity I saw was to vastly enhance the sponsor's ability to facilitate improved patient enrollment in clinical trials. The challenge was - how do we keep our company's clinical protocol on the top of the minds of the PI and SCs at a site? How do we ensure that no enrollable patient is missed? The sponsor of the trial is concerned with things like study design, site selection, study oversight and selected site interactions, but they clearly can't be everywhere on the ground all the time. Similarly, the CRA/monitor has responsibilities such as site activation, regulatory documentation management, data collection and clean up - but again, patient enrollment at sites is a challenge that is beyond the role and responsibility of CRAs. Therefore, a Clinical Trial

Liaison can be hired to fill some of the gaps at a local level. Their primary functions are to develop, foster, and leverage relationships with investigators, study coordinators, and key staff. Also, they can identify barriers to patient enrollment and provide customized solutions to those particular issues. They can maintain constant and local communication channels in order to motivate and energize key personnel. They can also provide protocol education and disease state awareness to all relevant personnel, including night and weekend staff. Finally, they can elevate the importance and value of a study relative to other protocols at a site. The efforts of a CTL can raise awareness of the study protocol, maintain motivation and engagement, reduce protocol violations and deviations, and ultimately increase study enrollment.

Just to clarify, what are the differences between a CTL and a CRA?

CRAs have primary responsibility for study site selection, qualification, and initiation in compliance with Good Clinical Practice (GCP) standards. The CTLs help facilitate these efforts through their relationships with sites and must work closely with CRAs to ensure that the processes are completed in a timely fashion. For example, site contracts can take 6 months to negotiate. Or, if properly managed, they can be done in less than a month. The timesavings are extremely valuable. CRAs are also responsible for ensuring accurate data collection by sites, and recovering source documentation in accordance with clinical operations protocol. This is very much regulatory and diligence activity. CRAs do NOT focus efforts on education and awareness activities at sites, on multi-shift staff education or weekend patient recruitment, for example. Also, each study site has a unique barrier to success, and this requires on-the-ground problem solving to identify unique solutions and ensure that they are implemented so that the site can be a successful study participant. CTLs are responsible for these activities. The skill sets required for these roles are quite different.

What skills do you seek when hiring CTLs and how are they different from MSLs?

It is my experience that about 2/3 of companies that employ MSLs require advanced scientific degrees. While this may be appropriate for the MSL role, I believe it is a mistake to make an advanced scientific degree a key requirement for CTLs. The objective of the CTL is the facilitation of clinical trials, and the skill sets required are different from those of CRAs or MSLs. The most important skills/qualities for CTLs are: relationship skills, problem solving skills, and scientific acumen, in this order. Thus, an advanced degree should not be the prerequisite for success as a CTL. Rather, I want a individual who can walk into a hospital and within hours find their way to all of the key stakeholders responsible for patient identification, screening, enrollment, and management. This might range from emergency room nurses to nurse educators, to pharmacists to sub-investigators, and the CTL has to form immediate relationships with all of these individuals and walk away with the ability to contact them for follow up education, motivation, etc. I also want someone who can recognize an obstacle and ask the right questions of the right people in order to find a solution and get it implemented. This may be as simple as addressing the lack of evening or weekend research coverage, or it may be as complex as a political rivalry between two investigators preventing access to a large bucket of patients for screening. Next, the CTL must have the ability to rapidly assimilate complex protocols, drug mechanism of action (MOA), etc. and then be able to educate staff members in an engaging manner, so scientific acumen is a requirement. I should also express the need for a strong work ethic, because the CTL can't leave a site until all of the issues listed above are addressed adequately. This may mean spending 2-3 days in a row at a single study site, or it could mean providing critical care unit in-services on the study and patient type at 10 PM on a Saturday night.

How can MSLs help clinical operations/research within a company?

MSLs can provide critical information to scientific thought leaders and study investigators, helping to keep them engaged with the sponsor and the study itself. MSLs can also field scientific research requests and facilitate sponsor-PI interaction to develop research proposals. However, to be very clear, I do NOT believe that MSLs can perform the same role that I described above as a "CTL". I do not believe that a MSL can serve two masters - there is a commercial role that MSLs fulfill, and there is a research role that CTLs fulfill, and I do not believe the two can be safely combined. I believe there is a case for both MSLs and CTLs to work for a company as separate and distinct functions.

What metrics or benchmarks have been helpful to you in discovering the value of MSLs working in concert with clinical operations, or what metrics can be used with CTLs?

Again, applying research metrics to MSLs is, I believe, inappropriate. However, with respect to CTLs, the metrics are numerous, and can include: patient enrollment per site, per month; number of protocol violations; number of protocol deviations; time to contract completion; time to institutional review board (IRB) approval; time to site initiation; time to first patient in at site; enrollment of appropriate versus inappropriate patients; etc. The ULTIMATE measurement of success in research is patient enrollment, and the ultimate goal is to reduce the study timeline.

Why should companies consider investment in a CTL group?

To save money, of course, and ultimately to get the drug to market in a shorter timeframe. An appropriately trained and directed CTL group can significantly reduce the timeline associated with any clinical trial through faster study start-up times and increased patient enrollment. Let's make up an example:

TREZZA BIOTECH* is conducting a 2,500 patient study that is planned to take 2.5 years to complete. While there is a ramp on the clinical operations budget according to study start up and growth in patient enrollment, let's assume that it is expected that the average monthly cost of the study will be $1.6M. However, things get off to a rocky start as they find their timelines lagging due to slow processing and turnaround of contracts, budgets, and IRB reviews, and they soon find that the study is already 2 months behind schedule and they haven't enrolled a patient yet.

Once they start enrolling patients, they also find that their enrollment projections were a bit aggressive and after 5 months of enrollment, they are well short of their initial objectives, and an additional 3 months behind timelines. As the study goes on, they continue to lag behind enrollment goals, and by the end of the study, the timeline was over-extended by a total of 8 months (2 months due to slow start up process and 6 months due to lagging enrollment).

Thus, in sum it may be estimated that the study required an additional $12.8M of investment. Further, because they are now 8 months delayed in drug development, they have lost 8 months of commercial life for the drug. Assume that the drug TREZZALINE*, is estimated to peak and plateau at $600M per year. This means that TREZZA will essentially lose 8 months of sales revenues at the end of the drug's patent life, and this equates to $400,000,000 in sales. From this perspective, EACH DAY saved through faster study enrollment would equal gains of $53K in near term expenses and $16.8M in down stream revenues. So the total cost of the delays in this study included both $12.8M in excess expenses associated with the prolonged clinical trial and $400M in lost sales during the course of commercialization.

From another perspective of this example, if CTLs successfully gain one month through increased enrollment, then they have reduced the clinical trial expense by approximately $1.6M and added one month to the commercial life of the drug, with an estimated value of $50M (undiscounted). Given that the cost of fielding a 5-member CTL team is approximately $1.4M, then it can be estimated that this group would pay for itself by simply improving enrollment enough to gain a few weeks.

I won't go on – we could discuss the costs and returns associated with CTLs for quite some time. But I will add that there are significant aspects of the CTL which may not be easily valued, but which I consider to be of inestimable value, such as: the relationships that are formed with the study PIs and staff – these are the early adopters who first use a therapy when it is commercialized; and the incredible knowledge that is attained through understanding of the complex patient management processes for any given disease state and how these individuals can transfer that knowledge to the commercial team preparing the drug for launch; or the understanding of the regional relationships between hospitals and key opinion leaders, and how these can be leveraged or addressed to support commercial launch. The list goes on, and you can see that as the study is completed and preparation for commercialization begins, CTLs have the opportunity to transition to a more MSL-like role, as the drug leaves the research arena and enters the commercial use arena.

What type(s) of training would a CTL need in order to produce?

CTLs must be trained on Good Clinical Practices, they must understand the roles and responsibilities of CRAs, Monitors, and Research Coordinators, and must understand the legal and regulatory implications associated with the conduct of clinical research. With respect to encouraging MSLs or other individuals to become CTLs, I would enlighten them to the wealth of knowledge that they would accumulate by experiencing clinical research from A-Z, present them with the opportunity to work with 'hard endpoints" (i.e. patient enrollment), and the challenges of conducting research at the site level, and prove to them that this experience would greatly enhance their careers and make them more marketable.

Have you seen any MSLs come out of a clinical operations/research role (such as a CRA role) and move into the MSL role?

If you can find the right individual with the right skills, this makes great sense - as these individuals LIVED the study, fully understand the patient type and the disease state, and have the closest experience with the drug that you will be commercializing.

Brian can be contacted at: brian.best@sbcglobal.net
                              510-338-0473

*Names in Brian's example were fictitious.

Future Of The MSL:
Closing Interview with Dr. Stan Bernard

We continue the discussion with Dr. Bernard regarding the future of MSLs from the past discussion at the beginning of the book. For Dr. Bernard's experience, please review the introduction section.

In your years of medical affairs work, what constitutes a 'great' MSL vs. an average MSL?

In my opinion, the good or great MSL is one that is perceived truly as a "peer" by other healthcare providers. MSLs need to earn clinical credibility with physicians and other healthcare practitioners. I think combining that with an ability to relate to their customers makes the good MSLs great.

What will future training look like for MSLs?

I see the future MSL transforming into a "health care professional consultant" representing pharmaceutical companies. They will be responsible for understanding new technologies as they emerge in the marketplace and communicating how they impact physicians, patients, and payers.

One example of an important emerging technology is pharmacogenomics: how patients respond differently to pharmaceuticals based on their genes. A future MSL role will be to help doctors and other healthcare professionals understand how and when to use pharmacogenomic testing and how to make clinical decisions based on the information provided. Because this technology is new to most physicians, the pharmaceutical industry will have to drive this education. Education provided by MSLs will become even more important given the increasing restrictions and lack of funding for medical education.

Another area of opportunity I see in the future for MSLs is interacting with other healthcare stakeholders besides physicians, such as policy-makers, media, pharmacists, nurses, managed care, and other important market influencers. MSLs will have an increasingly important role to play in educating and consulting with these constituents who have such a profound impact on product adoption and use.

We have spoken a lot about metrics in this book for the MSL. What metrics work best for MSL teams, in your opinion?

Demonstrating the value and benefits of the MSM program was perhaps our greatest challenge at BMS. Our first key metric was time spent with physicians. I believe that a key metric moving forward will be thought leader surveys and feedback. What do KOLs think about the quality of the interaction with MSLs? We have already seen third-party rankings of companies' sales force representatives and managed care account representatives. Over time we will see surveys for MSLs as well.

I think counting calls and related "reach and frequency models" are the wrong approaches to measuring MSL effectiveness. I've worked with some MSLs that had lots of calls, but not necessarily high quality calls. Reach and frequency are valuable metrics for sales representatives who need to deliver a certain number of marketing messages. However, if you impose these metrics on MSLs, the MSLs will begin acting like sales representatives. It is the quality of the interaction and how MSLs are interacting with their customers which result in the strong peer-to-peer relationships that are the foundation of effective MSL interactions.

Do you think the MSL role will endure?

I think the role will not only endure, but thrive. We are moving toward the golden age of MSLs. Sales representatives are declining, but the science is increasingly complex and will continuously evolve. Over time, MSLs will be recognized by pharmaceutical executives as a cornerstone for stakeholder education, communications, and relationship management.

In Conclusion

We hope that you have found this resource helpful in your decision to pursue, or not pursue a career as a Medical Science Liaison. Or, if you are already lucky enough to work in medical affairs, either as a MSL, a MSL manager, or someone linked to the MSL, we hope you learned at least one new idea on how to work better, smarter, or more efficiently. Like any job, there are good points and points for improvement within the MSL role.

We would love your feedback as well. Please write us at: info@msljobsatisfaction.com if you have further comments, questions and concerns. Finally, we wish you the best of luck with your career, whatever that may be, and we personally hope the MSL role not only endures, but evolves, in the months and years to come.

Appendix A:
Major Medical Associations/ Meetings for Therapeutic Areas

This list is not inclusive of all meetings, but gives some of the major medical meetings/associations within each therapeutic area as referenced in Chapter 1.

AIDS
- Interscience Conference On Antimicrobial Agents and Chemotherapy (ICAAC) - www.icaac.org
- International AIDS Conference (Every other even year) - www.aids2008.org

Cardiology
- American College of Cardiology (ACC) - www.acc.org
- American Heart Association (AHA) - www.americnaheart.org
- American Society of Hypertension (ASH) - www.ash-us.org
- Transcatheter Cardiovascular Therapeutics (TCT) - www.tct2007.com

Central Nervous System/Neuroscience/Neurology/Psychiatry
- Society for Neuroscience (SFN) - www.sfn.org
- American Academy of Neurology (AAN) - www.aan.com
- American College of Neuropharmacology (ACNP) - www.acnp.org

- College of Psychiatric and Neurologic Pharmacists (CPNP) - www.cpnp.org
- New Clinical Drug Evaluation Unit (NCDEU) - Meeting is co-sponsored by The National Institute of Mental Health and The American Society of Clinical Psychopharmacology - www.nimh.nih.gov/ncdeu/index.cfm
- American Psychiatric Association (APA) - www.psych.org
- Society of Biological Psychiatry (SOBP) - www.sobp.org

Critical Care/Emergency Medicine
- American College of Emergency Physicians (ACEP) - www.acep.org
- American Thoracic Society (ATS) - www.thoracic.org
- Society of Critical Care Medicine (SCCM) - www.sccm.org

Endocrinology
- American Association of Clinical Endocrinologists (AACE) - www.aace.com
- American Diabetes Association (ADA) - www.diabetes.org
- The Endocrine Society (ENDO) - www.endo-society.org
- The Metabolic Institute of America - www.insulinresistance.us
- American Thyroid Association (ATA) - www.thyroid.org
- Society for Endocrinology - www.endocrinology.org

Epidemiology
- American College of Epidemiology (ACE) - http://acepidemiology2.org/meetings/index.asp
- Society of Epidemiologic Research (SER) - www.epiresearch.org
- Society for Pediatric and Perinatal Epidemiologic Research (SPER) - www.sper.org
- Society for Healthcare Epidemiology of America (SHEA) - www.shea-online.org

Gastroentrology
- American College of Gastroenterology (ACG) - www.acg.gi.org

- American Society for Parenteral and Enteral Nutrition (ASPEN) - www.nutritioncare.org
- Digestive Disease Week (DDW) - www.ddw.org

General Medicine/Family Practice
- American Academy of Family Physicians (AAFP) - www.aafp.org
- American College of Physicians (ACP) - www.acponline.org
- American Medical Association (AMA) - www.ama-assn.org (This site also has a doctor locator - www.ama-assn.org/go/doctorfinder)
- Primed (East, West, Mid-Atlantic) - www.pri-med.com

Geriatrics
- American Association of Geriatric Psychiatry (AAGP) - www.aagpgpa.org
- American Geriatrics Society (AGS) - www.americangeriatrics.org
- Commission for Certification in Geriatric Pharmacy (CCGP) - www.ccgp.org
- Gerontological Society of America (GSA) - www.geron.org

Immunology
- American Association of Immunologists (AAI) - www.aai.org

Infectious Diseases
- Infectious Diseases Society of America (IDSA) - www.idsociety.org

OBGYN/Women's Health
- North American Menopause Society (NAMS) - www.menopause.org
- American Society for Reproductive Medicine (ASRM) - www.asrm.org
- American College of Gynecology (ACOG) - www.acog.org
- Association of Professors of Gynecology and Obstetrics and Council on Resident Education in Obstetrics and Gynecology (APGO/CREOG) - www.apgo.org

Ophthalmology
- American Academy of Ophthalmology (AAO) - www.aao.org
- American Academy of Optometry (AAOPT) - www.aaopt.org
- The Association for Research in Vision and Ophthalmology - www.arvo.org
- American Society of Cataract & Refractive Surgery (ASCRS) - www.ascrs.org
- Association of University Professors of Ophthalmology (AUPO) - www.aupo.org
- International Congress of Eye Research (ICER) - www.iser.org

Oncology/Hematology
- American Association for Cancer Research (AACR) - www.aacr.org
- American Cancer Society (ACS) - www.cancer.org
- American College of Radiology (ACR) - www.acr.org
- American Hepato-Pancreato-Billiary Association (AHPBA) - www.ahpba.org
- American Society for Therapeutic Radiology and Oncology (ASTRO) - www.astro.org
- Association of Community Cancer Centers (ACCC) - www.accc-cancer.org
- American Society of Breast Surgeons - www.breastsurgeons.org
- American Society of Clinical Oncology (ASCO) - www.asco.org
- American Society of Hematology (ASH) - www.hematology.org
- American Society of Pediatric Hematology/Oncology (ASPHO) - www.aspho.org
- Hematology Oncology Pharmacy Association (HOPA) - www.hoparx.org
- Oncology Nursing Society (ONS) - www.ons.org
- National Cancer Institute (NCI) - www.cancer.gov
- Radiological Society of North America (RSNA) - www.rsna.org
- Society of Interventional Radiology (SIR) - www.sirweb.org
- The Chemotherapy Foundation - http://mssm.edu/tcf/
- World Conference on Interventional Oncology (WCIO) - www.wcio2007.com

Osteoporosis/Bone Disorders
- American Society for Bone & Mineral Research (ASBMR) - www.asbmr.org
- International Bone & Mineral Society (IBMS) - www.ibmsonline.org
- International Conference on Children's Bone Health - www.iccbhh4.org
- International Osteoporosis Foundation (IOF) - www.iofbonehealth.org
- International Skeletal Society (ISS) - www.internationalskeletalsociety.com
- International Society of Clinical Densitometry (ISCD) - www.iscd.org
- National Osteoporosis Foundation (NOF) - www.nof.org

Over-The-Counter (OTC) drugs/Self Care
- American Pharmaceutical Association Self Care Institute - www.aphanet.org
- Nonprescription Medicines Academy (NMA) - www.nmafaculty.org

Pain
- American Pain Association (APA) - www.painassociation.org
- American Academy of Pain Medicine (AAPM) - www.painmed.org

Pediatrics
- American Academy of Pediatrics (AAP) - www.aap.org
- International Society for Pediatric and Adolescent Diabetes (ISPAD) - www.ispad.org

Pharmacy
- American Association of Colleges of Pharmacy (AACP) - www.aacp.org
- American College of Clinical Pharmacy (ACCP) - www.accp.com
- American College of Clinical Pharmacology (ACCP) - www.accp1.org

- American Institute of the History of Pharmacy (AIHP) - www.aihp.org
- Academy of Managed Care Pharmacy (AMCP) - www.amcp.org
- American Pharmaceutical Association (APhA) - www.aphanet.org
- American Society of Consultant Pharmacists (ASCP) - www.ascp.com
- American Society of Health System Pharmacists (ASHP) - www.ashp.org
- Board of Pharmaceutical Specialties (BPS) - www.bpsweb.org
- National Association of Chain Drug Stores (NACDS) - www.nacds.org
- National Community Pharmacists Association (NCPA) - www.ncpanet.org
- National Home Infusion Association (NHIA) - www.nhianet.org

Renal
- American Association of Kidney Patients (AAKP) - http://216.86.242.137/conventi.htm
- American Nephrology Nurses Association (ANNA) - http://anna.inurse.com/events/
- American Society of Nephrology (ASN) - www.asn-online.com
- International Society of Nephrology (ISN) - www.isn-online.org
- National Kidney Foundation (NKF) - www.kidney.org
- Polycystic Kidney Research Foundation (PKRF) - www.pkdcure.org
- Renal Association - www.renal.org
- Renal Physician's Association (RPA) - www.renalmd.org

Respiratory
- American Academy of Allergy, Asthma & Immunology (AAAAI) - www.aaaai.org
- American Association for Respiratory Care (AARC) - www.aarc.org
- American Thoracic Society (ATS) - www.thoracic.org
- American College of Chest Physicians (ACCP) CHEST is their annual meeting - www.chestnet.org

Rheumatology
- American College of Rheumatology (ACR) - www.rheumatology.org

Sleep
- Associated Professional Sleep Societies (APSS) - www.apss.org
- National Sleep Foundation (NSF) - www.sleepfoundation.org

Toxicology
- American Academy of Clinical Toxicology (AACT) - www.clintox.org
- American Association of Poison Control Centers (AAPCC) - www.aapcc.org
- American Board of Veterinary Toxicology (ABVT) - www.abvt.org
- American College of Medical Toxicology (ACMT) - www.acmt.net
- European Association of Poisons Centers and Clinical Toxicologists (EAPCCT) - www.eapcct.org
- Society of Toxicology (SOT) - www.toxicology.org

Transplantation
- American Society for Artificial Internal Organs (ASAIO) - http://asaio.com/annual.htm
- American Society of Transplantation (AST) - www.a-s-t.org
- American Society of Transplant Surgeons (ASTS) - www.asts.org
- The Transplantation Society - www.transplantation-soc.org
- United Network for Organ Sharing (UNOS) - www.unos.org

Urology
- American Urology Association (AUA) - www.auanet.org
- European Association of Urology (EAU) - www.euroweb.org

Other Industry Specific Organizations
- American Academy of Pharmaceutical Physicians (AAPP)
- American Association of Pharmaceutical Scientists (AAPS)
- American Society for Clinical Research Professionals (ASCPT)
- Association of Clinical Research Professionals (ACRP)

- Biotechnology Industry Association (BIO)
- Drug Information Association (DIA)
- Food and Drug Law Institute (FDLI)
- International Society for Pharmacoepidemiology (ISPE)
- Pharmaceutical Education and Research Institute (PERI)
- Pharmaceutical Researchers and Manufacturers of America (PhRMA)
- Regulatory Affairs Professional Society (RAPS)
- Society of Clinical Research Associates (SOCRA)
- Society for Clinical Data Management (SCDM)
- The Organization for Professionals Regulatory Affairs (TOPRA)

MSL Specific Sites, Companies and Discussion Boards
- Pharm, LLC - www.pharmllc.com
- Every first Friday in November is MSL Appreciation Day. You can find this holiday listing in the *2008 Chase's Calendar of Events*.
- Cafe Pharma, MSL board - www.cafepharma.com/boards/forumdisplay.php?f=75
- MSL Institute - www.mslinstitute.com
- MSL Forums.com - http://forums.mslforums.com/forums/default.aspx
- MDea - www.mdeany.com
- The Medical Affairs Company - www.themedicalaffairscompany.com
- Science Oriented Solutions - www.medicalaffairs.com
- Scientific Advantage - www.scientificadvantage.com

Appendix B: Real World Career Paths to and Beyond the MSL

In this section, there are 10 real world career paths from 10 different professional background types including the path of the field based MSL. These are provided in hopes to demonstrate the variety of professional backgrounds that can lead to the MSL role.

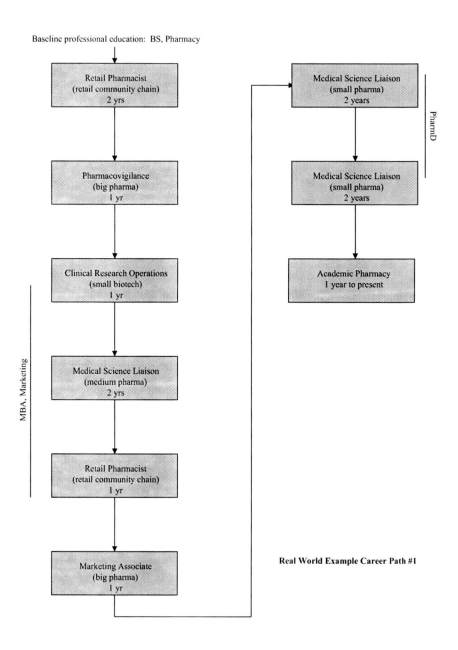

Real World Example Career Path #1

The Medical Science Liaison

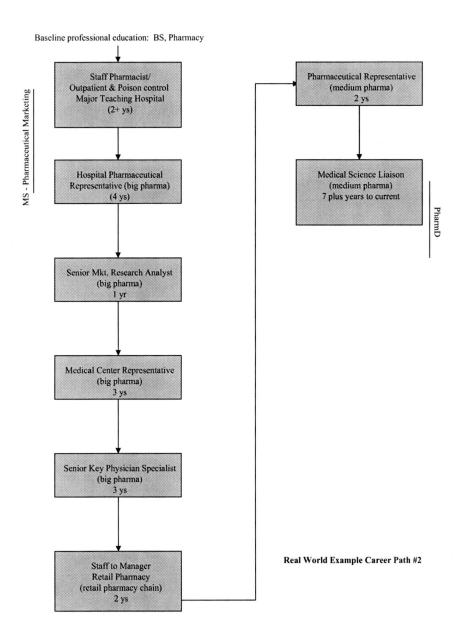

Real World Example Career Path #2

Baseline professional education: BA, Business Administration

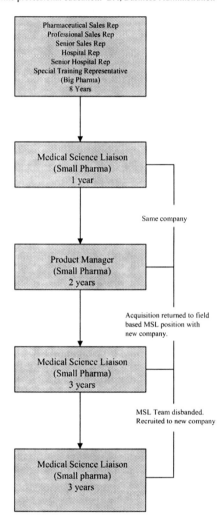

Real World Example Career Path #3

The Medical Science Liaison

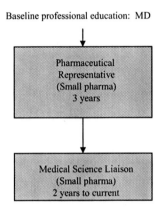

Real World Example Career Path #4

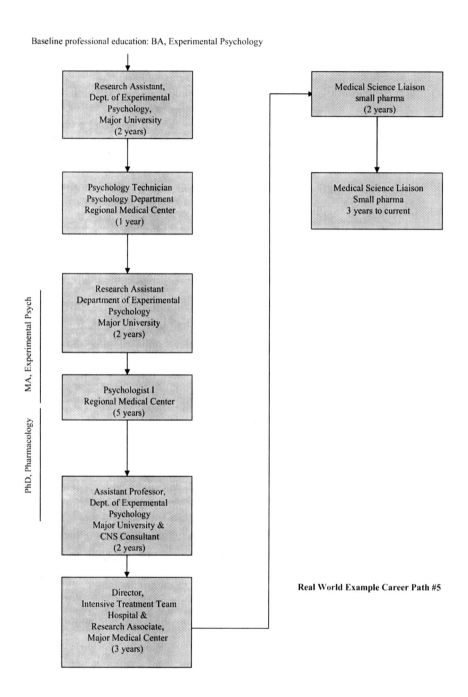

Real World Example Career Path #5

The Medical Science Liaison

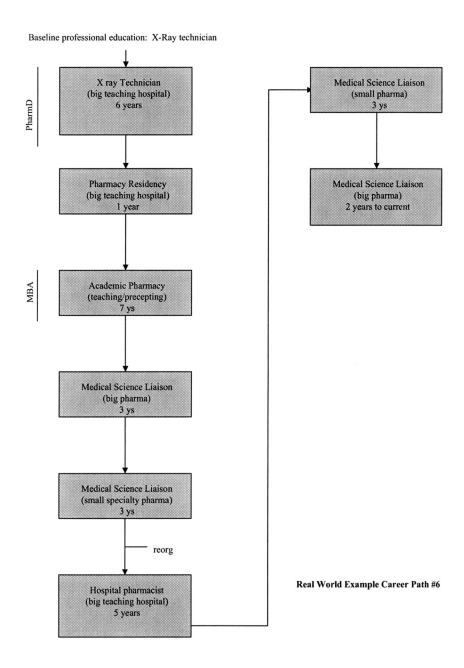

Real World Example Career Path #6

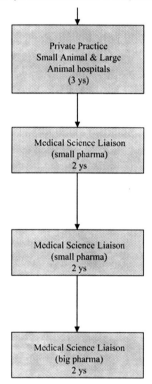

Real World Example Career Path #7

The Medical Science Liaison

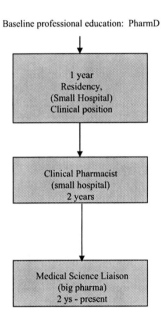

Real World Example Career Path #8

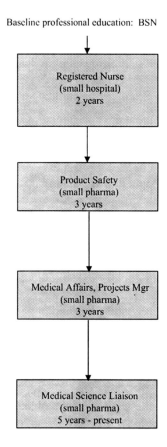

Real World Example Career Path #9

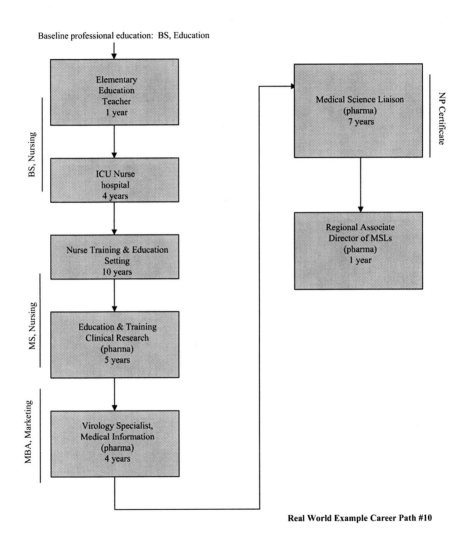

Real World Example Career Path #10

Appendix C:
Acronyms & Abbreviations

A listing of all abbreviations used in this handbook is included below for reference.

AACP - American Association of Colleges of Pharmacy

ACCME - American Council on Continuing Medical Education

ACCP - The American College of Clinical Pharmacy Accp.com.

ACPE - American Council on Pharmacy Education

ADME - Absorption, distribution, metabolism, and excretion of a compound or drug through the body. Also known as pharmacokinetics, which is the study of ADME.

AdvaMED - Advanced Medical Technology Association. Organization for medical device manufacturers.

AHA - American Heart Association

AJPE - American Journal of Pharmaceutical Education

AMA - American Medical Association

APA - American Psychiatric Association

APhA - The American Pharmaceutical Association. Aphanet.org.

ASCO - American Society of Clinical Oncology

ASRT - American Society of Radiologic Technologists. asrt.org

BA - Bachelor's of the Arts

BCNP - Board certified nuclear pharmacist

BCNSP - Board certified nutrition support pharmacist

BCOP - Board certified oncology pharmacist

BCPP - Board certified psychiatric pharmacist

BCPS - Board certified pharmacotherapy specialist

BS - Bachelor of Science Degree

BSN - Bachelor of Science in Nursing

CBI - Center for Business Intelligence. For profit, private company that holds professional industry wide meetings, some of which are MSL specific. See cbinet.com for information.

CCD - Certified Clinical Densitometrist

CCRC - Certified Clinical Research Coordinator. socra.org has more information.

CDC - Centers for Disease Control

CDE - Certified Diabetes Educator. To learn more, logon to ncbde.org.

CDER - The Center for Drug Evaluation and Research

CE - Continuing Education

CEC - Clinical Education Coordinator

CEO - Chief Executive Officer

CI - 1. Competitive Intelligence. This is the process of ethically gathering information on competitor companies or products. One can become accredited as a certified competitive intelligence professional (CIP) through SCIP (Society of Competitive Intelligence Professionals) www.scip.org. 2. Confidence Interval.

CIP - Certified Competitive Intelligence Professional

CME - Continuing Medical Education

CMO - Contract Medical Organization

CRA - Clinical Research Associate

CRM - Customer relationship management

CRO - Contract Research Organization

CSAM - Certified Sales Account Manager

CSL - Clinical Science Liaison - sometimes, the same thing as a MSL. Other companies have the CSL provide more clinical information as a nurse educator.

CTL - Clinical Trial Liaison

CV - curriculum vitae. A more detailed résumé that includes lectures, posters, and publications.

DABAT - Diplomat of the American Board of Applied Toxicology: board certification for health care professionals in toxicology. www.clintox.org.

DBSA - Depression & Bipolar Support Alliance

DDMAC - The Division of Drug Marketing Advertising and Communications. Part of FDA that regulates pharmaceutical promotion and advertising, and a branch of the Center for Drug Evaluation and Research (CDER). Visit http://www.fda.gov/cder/ddmac/ for more information.

DEA - Drug Enforcement Agency.

DIA - The Drug Information Association. Membership based organization for professionals in drug information and the pharmaceutical industry. Logon to www.diahome.org for information.

DIJ - *The Drug Information Journal*. Publication of The Drug Information Association, or DIA.

DOJ - Department of Justice

DVM - Doctor of Veterinary Medicine

EBSCO - publishing company with a database called EBSCOhost®

EdD - Doctor of education

EMBASE™ - Suite of products/databases from Elsevier, a publisher.

Esq. - Esquire or attorney

Etoc - Electronic table of content. New England Journal of Medicine, for example, sends out an Etoc previewing their new magazine content for free.

EU - European Union

FACOG - Fellow of the American College of Obstetricians and Gynecologists

FDA - Food and Drug Administration. Visit www.fda.gov for more information.

FDAMA - The Food and Drug Administration Modernization Act

FTE - Full time employee

GCPs - Good Clinical Practices: Guidelines on how to run clinical trials. From International Conference on Harmonisation (ICH).

GLPs - Good Laboratory Practices

GPS - Global Positioning System

HBA - The Healthcare Business Women's Association. A not for profit membership based organization promoting women in healthcare. For more information, logon to www.hbanet.org.

HCP - Healthcare professional

HECON - Health economics

HHS - The United States Department of Health & Human Services.

HIT - Health Information Technology: Emerging term for scientists working in technology based applications in order to perform research regarding medicine.

HMO - Health Maintenance Organization in managed care

HR - Human Resources

ICH - International Conference on Harmonisation: Visit www.ich.org for more information.

ICSD - International Classification for Sleep Disorders

IDIS - Iowa Drug Information Service

IIR - Investigator-initiated Research: See IIT.

IIT - Investigator-initiated Trial: A study protocol or concept conceived by an investigator, rather than by a drug company. Also known as investigator-initiated research (IIR).

IPA - International Pharmaceutical Abstracts

IQPC - Company that holds some conferences relating to the MSL's work. www.iqpc.com.

IRB - Institutional Review Board

ISCD - International Society for Clinical Densitometry

ITT - Intent to treat

KOL - Key Opinion Leader: Another term for academic thought leader or thought leader, which is usually the primary customer of the MSL. Also known as a thought leader (TL) or key thought leader (KTL).

L&A - Licensing & Acquisition

LOCF - Last observation carried forward

LTC - Long term care

MA - Master's of the Arts

MBA - Master's of Business Administration

MD - Doctor of Medicine

MeSH terms - Medical subject headings. Articles indexed in Medline are indexed via medical subject headings.

MOA - Mechanism of action

MPA - Master's of Public Administration

MPH - Master's of Public Health

MS - Master's of Science

MSL - Medical Science Liaison

MSM - Medical Service Manager: Squibb's first scientist/business based medical science liaison professionals. They did not use the term MSL because they were attempting to differentiate from Upjohn's MSL team, which came primarily from a sales background rather than a science background.

NAMI - National Alliance on Mental Illness

NAMS - The North American Menopause Society

NCBI - National Center for Biotechnology Information

NCDEU - New Clinical Drug Evaluation Unit

NDMDA - National Depressive and Manic-Depressive Association

NIH - The National Institutes of Health. www.nih.gov.

NLM - National Library of Medicine

NNT - Number needed to treat

NP - Nurse Practitioner

OIG - Office of the Inspector General. Part of the United States Department of Health & Human Services (HHS). Logon to www.hhs.oig.gov for details.

OTC - Over-The-Counter: medications sold over the counter.

OUS - Outside the United States

OVID® - a database

PDA - Personal digital assistant

PDR - Physician's Desk Reference

PharmD - Doctor of Pharmacy or doctorate of pharmacy

Phase I, II, III, IIIB, or IV research - Different phases of clinical research required by the government for most investigational compounds, biologicals, drugs, or devices in order to seek approval by the Food and Drug Administration. For more information on the clinical trial process, logon to www.clinicaltrials.gov and click on 'understanding trials'.

PhD - Doctor of Philosophy

PhRMA - The Pharmaceutical Research and Manufacturers of America

PI - 1. Principal Investigator: The primary physician working on a protocol or clinical trial at a specific clinical study site in relation to clinical research. 2. Package Insert: The official product information on a drug. Also known as a package label or a product label.

PK - Pharmacokinetics, or the study of absorption, distribution, metabolism, and excretion of a drug or compound through the body.

PPO - Preferred Provider Organization in managed care

PRN - as needed

PubMed - Service of the US National Library of Medicine, which indexes articles from Medline and other sources. www.pubmed.gov.

R&D - Research and Development

RD - Registered Dietician

RN - Registered Nurse

ROI - return on investment

RPh - Registered Pharmacist

RSUs - restricted stock options

SC - Study Coordinator

SFF - Site Feasibility Forms: questionnaires produced by companies to gather and identify clinical research centers for company sponsored research trials.

SOPs - Standard Operating Procedures in clinical trials.

Toxnet - Part of the National Library of Medicine. Primarily drug safety data. http://toxnet.nlm.nih.gov.

SIAC - Special Interest Area Communities. Smaller specialty areas of the Drug Information Association.

TL - Thought leader. See KOL for definition.

UEG - unrestricted educational grant

WLF - The Washington Legal Foundation

Bibliography, References, and Resources

1. Carlisle, R. *A Century of Caring: The Upjohn Story.* Elms Ford, NY: Benjamin Co; 1987: 144-146.

2. Morgan D, Domann D, Collins E, Massey K, Moss R. History and Evolution of field-based Medical Programs. *Drug Information Journal.* 2000: 34; 1049-1052.

3. First year MSL job satisfaction survey data: 2003. Albert E, Sass C. Medical Liaison Job Satisfaction. *Medical Science Liaison Quarterly.* March 15, 2004: 2 (1).

4. Herzberg F. One More Time: How Do You Motivate Employees? *Harvard Business Review.* January, 2003: Best of HBR. Article R0301F.

5. Second year MSL job satisfaction survey data: 2004. Albert E, Sass C. Field-Based Medical Science Liaison Job Satisfaction. Poster and presentation at the *Drug Information Association Annual Medical Communication Workshop.* March, 2004.

6. Third year MSL job satisfaction survey data: 2005. Albert E, Sass C. Field-based Medical Science Liaison Job Satisfaction: Three-Year Results. Poster #335 presented at the 2005 Annual Meeting of the American College of Clinical Pharmacy, October 2006.

7. Fourth year MSL job satisfaction survey data: Albert E, Sass C. A Survey of Medical Science Liaisons. *Pharmaceutical Executive*. October 2006: 64-68. Available online at: http://www.pharmexec.com/pharmexec/article/articleDetail.jsp?id=378063&searchString=Medical%20liaisons

8. Fifth year MSL job satisfaction survey data: Data on file, Pharm, LLC.

9. Buckingham M, Clifton D. *Now, Discover Your Strengths*. New York: The Free Press; 2001.

10. Yate M. *Cover Letters That Knock 'em Dead*. Avon, MA: Adams Media; 2006: 12-13.

11. When you think you are having a bad day. Picture available online at: http://darwin.chem.villanova.edu/~bausch/images/badday1.jpg.

12. Emoto M. *The Hidden Messages in Water*. Hillsborough, OR: Beyond Words Publishing Inc; 2004.

13. Jung CG. *The Archetypes and the Collective Unconscious*. Princeton, NJ: Princeton University Press; 1990.

14. Bohm D. *An Interview with David Bohm*. Videotape. New York: Mystic Fire Video, Inc. 1994.

15. Phillips D. *Lincoln on Leadership: Executive Strategies for Tough Times*. New York: Warner Books; 1992.

16. The American College of Clinical Pharmacy has a publication on residencies and fellowships available in the US. For more information, visit www.accp.com.

17. Vizcaino versus Microsoft® Corp., 1995, 1999. Located online at: http://www.techlawjournal.com/courts/vizcaino/19990512.htm and lay press summary located online at: http://www.washtech.org/news/courts/display.php?ID_Content=381.

18. Accountemps Survey - story located online at: www.insideindianabusiness.com/newsitem.asp?id=24876.

Acknowledgements

This book could not have been possible without the support and encouragement of many others. They include:

- Everyone that graciously agreed to interview for this book project, and those that donated career paths. Without them, this book could not have been published.
- All our former and current co-workers, colleagues, managers, and directors.
- All the people that inspire us to try new things.
- Our book mentor and friend Elaine Voci. Without her this journey would not have been possible.
- Chris Russell, guardian of our crazy ideas.
- To our friends and family, who supported the project in any way that they could.
- To Butler University, Dr. Julie Koehler, and Dean Mary Andritz for their support of this project.
- Most importantly, to all the field-based medical science liaisons out there, working to make the profession of the MSL one of the greatest jobs on the planet.